c/95

AF835258

j Milne
Gift from the sky
cop.1

JAN 31 1968

Highland Park Public Library
Highland Park, Illinois
ID 2-0216

Gift from the Sky

GIFT FROM THE SKY

Lorus & Margery Milne

WITHDRAWN

ATHENEUM 1967 NEW YORK

Highland Park Public Library

J

COP. 1

Photographs of South American and Australian swans courtesy of Caribbean Gardens, Naples, Florida. Jacket photograph and photograph of Alice in late fall are by the authors. All other photographs courtesy of L. Franklin Heald.

Copyright © 1967 by Lorus J. Milne and Margery Milne
All rights reserved
Library of Congress Catalog card number 67-19006
Published simultaneously in Canada by
McClelland & Stewart Ltd.
Manufactured in the United States of America
Printed by Halliday Lithograph Corporation
West Hanover, Massachusetts
Bound by H. Wolff, New York
Designed by David Rogers
First Edition

To Brad, the magistrate who gave the first dollar; to Dave, the professor who tackled each problem successfully; and to Howard, who went to feed her every day; we dedicate this story of a wild white swan that changed the life of a little town and brought beauty to a quiet backwater in New England.

Contents

Gift from the Sky

Gift from the Sky

THE WEATHER MAN predicted the change correctly, but not the gift that came with it from the sky. Late at night, the cold northern air that had chilled the land with the tingle of autumn, blew out to sea. Warmth from the South began flowing up along the coast. Summer was coming back again. This was Indian summer, the kind of October weather that always catches people unawares.

More than an hour before sunrise, the clear sky brightened. Most of the stars faded for the day, but Jupiter still shone as the brilliant morning star.

A flock of Canada geese had enough light to see the coastline clearly. In wedge formation they flew, honking companionably to one another. It was bright enough, too, for a solitary swan that was trying to keep up with the fast-flying geese. She had taken off with them half an hour before, soon after twilight began. Now her wing muscles were beginning to ache from so much exercise. As she tired, she looked below for a place to rest.

Far ahead, she saw a great triangular expanse of open water, surrounded on all sides by dark fields and forests. The big white swan slowed her flight a little, and slanted down toward the great inland bay. The geese flew on, and were soon lost to sight.

The swan settled on the nearest inlet. She splashed down heavily, dipped her pink beak into the water, and let some of the cool liquid slide down inside her throat. It was slightly salty. This was an estuary, where several small rivers joined together before running into the sea. Whenever the tide was high, salt water flowed in and gave its flavor to the fresh water.

Arching her neck into a graceful S-curve, the swan turned westward. She floated high on the dark water as she pushed herself along, kicking steadily with one big webbed foot, letting the other one trail. Behind her, beyond the hill to the east, the red rim of the rising sun came up. It silhouetted the trees and brightened the sky still more. Where the swan swam, the tide was almost full. The water scarcely moved. Over its glistening surface, mist rose in silvery spirals back-lit by the rising run. The wind stopped blowing, and silence spread everywhere.

Farther up the estuary the mist concealed an old dam, which held back the waters of the nearest river. It produced a quiet pond that few people even noticed any more. The swan was too far away to hear the sound of water from the pond pouring over the old dam into the estuary.

Because this was autumn, the swan propelled herself through a world of color. She could see the leaves hanging golden on the poplars, rich red-brown on the oaks, cinnamon-colored on the beeches, and scarlet on the maples. The elms had none at all, for theirs lay flat on the ground. Against the misty sky at sunrise, their bare branches spread blackly, tall and still.

Sparkling ice crystals coated the surface of the ground, showing how cold the early part of the night had been. Like frosting they covered many of the dried goldenrods and the milkweeds on low land near the water. The pods on the milkweeds had split just enough to reveal their silky-tufted seeds. They glistened, white as the swan. With a few more sunny days, those seeds would be parachuting on the breeze.

Farther down the estuary, two duck hunters shivered slightly inside their heavy jackets and thin gloves. Seated in their skiff, they were hidden among a camouflage they had woven with leafy branches. The men turned their heads slowly from side to side, but made no other move. They hoped that if a flight of ducks came near, the birds would not recognize the steamy breath that condensed like a little cloud above the skiff every time a man exhaled. Anyway, no ducks came.

Then the hunter who had his back to the sun saw something white. It shone on the water close to the shore half a mile away. "Hey! What's that?" he whispered to his friend. "Look away up, on the right-hand side. Something big and white."

The other hunter glanced over his shoulder, but saw nothing. "Is it a bird?"

"I think so."

"If it's big and white, it's no duck. Forget it!"

"It might be a snow goose."

"Closed season here on snow geese. Probably somebody's barnyard goose. Like I say, forget it."

"We're not doing so well with ducks. Why not give up and go see what that white bird is?"

"Let's wait a little longer."

So the two men sat, while the sun moved ever higher in the sky. It dispelled the mists and followed the wan semicircle of the moon across the blue. The air warmed gradually, and the hunters loosened their jackets. Finally, the reluctant one said, "Okay. Let's go."

The other man pulled the starter cord twice on the outboard motor before it coughed to life. He threw the machine into gear and steered the skiff up the estuary toward the place where he had seen the white bird. It had vanished from sight, but he still expected to find it. He felt sure that no goose so large and white could fly off without being seen. But as the skiff sped up the estuary, he grew less sure. Could they have passed the bird while it hid among the marsh grasses? How could he have seen a goose so far away? It must have been closer to have appeared so big.

His friend sat squarely in the bow of the skiff, facing forward. The man in the stern could not tell from the back of his companion's cap, or his neck, or his broad back in its heavy jacket whether his eyes were closed for a snooze, or were searching the shoreline ahead. Yet the man's hand gripped the shotgun in a way that could bring it into firing position in a fraction of a second if any target appeared. Then he clearly gave up. He turned the gun crosswise, broke it, took out the shells, and snapped it shut again. Facing as far as he could toward the stern, he called back in a voice that carried above the noise of the outboard motor: "Your bird's gone. Too bad!"

The man's voice, the roar of the outboard motor, a flash of sunlight reflected from the barrel of the shotgun, and the onrushing boat all burst at once into the quiet backwater. There the swan had found some green seaweeds to eat as the tide ebbed out of a little gutter in the marshy edge of the estuary. Startled, the bird extended her magnificent wings high over her back and swept them outward and downward. Compressing the air below her wings, she lifted her heavy body out of the water in one swift move. Again and again she flapped, while with her left leg she kicked at the water surface with a webbed foot, helping herself gain forward speed.

To the men, the white bird seemed to explode out of the gutter mouth. Never before had they seen anything so huge borne away from them on feathery wings. The swan's long neck stretched out straight ahead. Soon its feet trailed behind, beside its short tail. Flying furiously, the bird attained its top speed of fifty miles an hour in just a few seconds. It fairly shot up the estuary, rose high over the old dam, and vanished from sight beyond the trees. Above the dam lay the broad millpond, on the edge of the town, where no hunting was allowed. The swan was safe.

The hunters turned back. They headed directly now for the distant ramp where they had left their boat trailer and automobile. The man who steered looked as though he had seen an angel.

"That was a swan, man! I told you it was a big bird. But I didn't expect it to be that big! It must have weighed close to forty pounds. That's twice as much as the biggest wild goose."

On the Millpond

As the frightened swan zoomed over the dam, she saw more people. Three boys sat fishing on the flat bottom of a homemade rowboat. It floated just above the dam, tied to a willow branch that overhung the pond. The two youngest boys wore brilliant orange life jackets. The thirteen-year-old had none to wear, but he was stronger and could swim well. Actually, the boat was so heavy that all three could lean over one side with no danger of its tipping.

Quickly the swan swerved to avoid colliding with the boat, the boys, and their long fishing poles. She sped away to circle the pond, looking for a good place to rest again. Her bright black eyes noticed every little bay and point of land. She noticed the darker water where the pond was deep, and the many shallow places where green plants grew close beneath the surface. She returned to the middle of the deep part, checked her forward speed, and fluttered to a stop on the water.

The boys almost jumped out of their boat as the huge white bird came straight at them, then veered off to pass only a pole's length away. The swan seemed bigger and faster than the jet airplanes from the air base. Its beating wings gave off a singing musical note, a throbbing sound unlike any other. At such close range, the boys' eyes gained only a blurred picture of the bird itself. Even when it settled on the pond a quarter of a mile away, they hardly knew what they had seen.

Their courage soon recovered. The boldest boy cast off from the willow branch. He seized the oars and began to row toward the swan. The other two reeled in the fishing lines and cleared the boat for action. This was far more exciting than waiting for a fish. They had been on the pond since sunrise, because their grandfather had told them the fish bit oftener before breakfast. Now they forgot their dreams of catching a salmon or something big. They shoved aside the perch and sunfish that flopped hopelessly on the bottom planking of the boat, and thought only of getting close to the huge white bird.

For a few minutes, nothing happened to alarm the swan. She sampled the water in the pond and found it fresh, with none of the salty flavor below the dam. She headed for a shallow place where the tips of some plants poked up through the water. They looked good to eat. But she kept an eye on the boys in the boat. She felt safer when people stayed on land and left the water to water birds.

The heavy boat was moving, but not coming any closer. It made no obvious headway because it was crossing the part of the pound where the current showed the most strength. The water moved the craft toward the dam almost as fast as the lone oarsman could propel it upstream. Gradually, however, he reached the quiet shallows near shore and escaped the current. Hauling then with all his strength, he speeded up.

The swan turned quickly from her pursuers and began swimming. Though she held her right leg stiffly to the rear, she made good progress with her left. She glided along almost as fast as the boat. But slowly the boys closed the gap between them. As the distance lessened, the swan raised her wings in readiness to fly. Then she relaxed and swam on. The boat had stopped.

The oldest boy, who was rowing, faced backward and had to depend upon the two younger ones to guide him. They had managed perfectly with "More to this side" or "More to that side." But they failed to warn him of a series of plants growing up through the water in the path of the boat. Now the boat was firmly aground on a mudbank, the plants crushed beneath the bottom planking. All of the interest the boys had had in catching up with the swan was soon drowned in the sweat of pushing with the oars against the mud.

While the boys struggled to get afloat again, the swan sailed to a safe distance and began to feed. She found some delicious plants and plucked off the freshest leaves one or two at a time. A lot of these greens were needed to satisfy her hunger. She was such a large bird.

Nothing so huge and white could long be overlooked. As a matter of fact, while she was hurrying away from the boys, a salesman of storm doors and double windows noticed her. He drew his car off the side of the road, close to the pond, and rolled down the glass so that he could see better. When he realized that the boys had given up chasing the strange bird, he drove home and telephoned excitedly to a naturalist friend.

"There's a huge white bird on the pond," he said. "It looks like a swan to me. It's far too big to be a goose. I wish you'd go and look at it."

The naturalist got his bird glasses and a camera with a telephoto lens, and hurried down to the pond. Any kind of a swan on the millpond would be news for the local bird-watchers. The only wild swans in the eastern states are whistling swans, which raise their families north of the Arctic Circle. In October and early November they travel south with their youngsters, to winter in salt marshes and coastal lakes from Massachusetts to northern Florida. Great flocks of them spend the cold months on a man-made marsh in North Carolina, where they are protected in a national wildlife refuge. But none had ever been seen before on the estuary in New Hampshire. Perhaps this bird was injured and had settled on the pond to rest.

Standing on the exact spot where the salesman had stopped his car, the naturalist could see the swan at the far side of the pond. Through his powerful glasses, he noticed that the bird's right leg was stretched out straight behind, just above the water, as it moved from place to place. Probably the leg was injured. He could not guess how badly.

One fact puzzled him: this was no whistling swan. It held its neck in a graceful curve, like the swans of literature, not stiffly upright as a whistler does. Its wings were conspicuous above its back, not part of a smooth contour, as they would have been had the bird been a whistler or any of the geese. Moreover, its beak seemed mostly a pinkish orange, whereas that of a whistler is black with a yellow streak below and in front of the eye. Surely it was a domestic swan—the royal bird of Europe. Where did it come from, and to whom did it belong? Domestic swans always belong to somebody!

To be sure of his identification, the naturalist walked all the way around to the far side of the pond. He approached the water from behind some bushes and rocks, for he did not want to alarm the bird. When, at last, he rose up within sight of the swan to take another photograph, the beautiful creature showed no real fear. It turned this way and that, arching its neck to nibble at plants in the shallow water.

The swan was really enjoying herself. Yet she

didn't forget for a minute that she had company. Every fifteen or twenty seconds, she lifted her head and looked around, just to make sure that she was safe. Between times she fed hungrily on plantain and other pondweeds. Whenever she raised her beak out of the water, it was full of dripping greenery. Sometimes a streamer of bright green algae hung there for a moment, until she shook her head to toss it off. From all sides of her, ripples spread in every direction.

She noticed someone else approaching the pond, striding down the sloping field between the low clumps of dogwood and occasional spruce trees. The naturalist recognized the newcomer to be the owner of the land—one of the three selectmen in the little town. Using a soft voice that he hoped would not alarm the bird, the naturalist called across: "Come slowly. There's an injured swan right ahead of you, close to shore."

The owner nodded and circled around to join the naturalist. "How did you know the bird was here?"

The naturalist told him of the sharp-eyed window salesman.

"I was down at the launching ramp on the estuary," the owner said. "A couple of duck hunters came in with their boat. They told me they had flushed a swan. It got out of range in a hurry and flew over the dam into the pond. If they'd told me they'd seen an ostrich, I couldn't have been more doubtful. But I was coming home anyway. So I decided to walk down here to check."

As the two men talked in low tones, the swan listened to their voices. Somehow the sound had a soothing quality. For more than a thousand years her ancestors had grown up where human speech was as familiar as bird songs. Far back, her kind had been widespread in Europe. Today, only in the lake district of Poland can large nesting colonies of wild ones still be found. But there are many tame ones, for Europeans adopted some of the Polish swans during the tenth century, and began raising them on large estates.

Because the commonest sound these birds make is a loud hiss, the name "mute swan" was given to them by the English. The French call the bird a *cygne muet.* The Dutch say *Knobbelzwaan,* the Swedes a *Knölsvan,* and the Germans a *Höckerschwan.* These three all mean the same—a knobbed swan. They refer to a black knob where the top of the beak meets the head. That knob grows very prominent, particularly on a male in breeding season. No other kind of swan has such a knob.

The knob was not very big on the swan that swam in the millpond. She was a young bird, and it had not yet grown to full size. Her beak was still pinkish, although not the pastel pink it had been when she was younger. Soon it would be the clear orange of maturity.

"Is this the kind of swan we see in public parks?"

"Yes," said the naturalist. "But park swans are usually prevented from flying away. The caretakers clip the flight feathers on one wing, or a veterinarian removes one pinion—the thumblike part of the wing on which the longest feathers grow. A pinioned bird can never fly again. One with a clipped wing gets new feathers when it molts, and will fly unless it is clipped regularly. A young bird, raised in captivity, may be left unclipped or unpinioned until it can fly. This one may simply have taken off on its own."

The swan nodded her head, as though in complete agreement, dislodging another long streamer of green algae. She was finding the pond wonderfully full of food, a banquet of all the things she liked best. Until recent rains had filled the pond, it had been choked with vegetation. But now, with the water level high, all of the greenery was nicely covered with water and spread apart, helping her reach it easily.

"Have you any idea where this swan came from?"

The naturalist shrugged his shoulders. "Who can say?" he replied. "The bird may have come from some estate within twenty miles. Or it may have come all the way from Europe. You remember that big storm we had last week, when the east wind kept up for hours? Traveling in such a storm, aided by an east wind, a swan might need only about thirty hours of flying time to get here from England."

"That's too long without a rest."

"Yes," the naturalist agreed. "And the storm probably wouldn't cover the entire ocean. But the bird could rest on the open water and, if it stayed in one of the ocean-going ship lanes, it might even find something to eat in the garbage that's thrown overboard. Though probably it could go two days without food, which would be plenty of time for such a trip."

The swan held her head high, straightening her neck and peering across the pond. She was trying to see what had made two loud banging sounds that came unexpectedly across the open water. They were caused by the slamming of two automobile doors; two cars had stopped on the far shore of the pond. Assorted children and adults had jumped out and hurried to the water's edge. Some of them were pointing to the large white swan shining in the sunlight.

"It would be a good thing to have swans on the town pond," the land owner commented, talking as a selectman now. "What would we do with them, though, when the pond freezes over in winter?"

"I don't know," the naturalist admitted. "Wild swans go into salt water along the coast, or migrate south far enough that the ponds stay open all winter. If this swan stays, I'll find out about housing and feeding it through the cold months. For the present, it's enough to see that the bird recovers and to find out if we can make friends with it."

Have You Seen the Wild Swan?

WHEN THE SKY above the pond grew dark that Sunday evening, the swan swam across the pond to the side where the road came close. For nearly an hour no cars had stopped. The few that passed were just dark shapes with a humming sound and lights that did not shine directly on the pond. The swan found so many delicious waterweeds to eat close to the shore that she had no time to notice the naturalist sitting there, his bird glasses in his hand. The naturalist had seldom felt happier. To be overlooked by a wild swan, and to hear it gulping noisily at vegetation in the water just a few feet away was a pleasure he had never expected to enjoy.

The swan spent the night among the low plants that grew out of the water near the road. She paid no attention to the small fishes that occasionally jumped and splashed in the moonlight. She ignored the black skunk with a white stripe down its back, rustling among the dead leaves on shore in search of worms and insects. The continuous trilling of the tree crickets, like elfin bells in the darkness, meant nothing to the swan. Neither did the squeaking of bats as they flew overhead, or the hoot of an owl, or the skittering sounds of flying squirrels frisking about on the bark of trees. None of them could harm the big bird.

At daybreak, the swan raised her head as high as possible and looked in all directions, studying her surroundings. She watched a pair of crows fly across the pond. A pileated woodpecker, almost as big as a crow but with a bright red crest, came out of the woodland near shore. He flashed the white on his wings, and followed an undulating course as though he were swooping along on an invisible roller coaster.

The swan began to preen her feathers, going over them one by one, dressing them in her beak. She had about four thousand feathers on her body alone, and nearly four times that many on her long neck and head. She needed hours every day to groom all those feathers. She had not nearly finished her job when the first rays of the rising sun shone on her.

An early motorist on the road saw the bird and stopped. He had never seen a swan on the pond before. As soon as the man got out of his car for a closer look, the swan swam away. The motorist went back to his car and drove off. The swan traveled to the far side of the pond and found more to eat there. The naturalist watched the bird go, satisfied that again it had found a safe place for the day.

But even across the wide pond, the big white bird attracted attention. An old car stopped and several men got out of it to stare at the swan. Another two cars pulled up, their drivers curious to learn why the men were standing there. The cars blocked the road for a building contractor who was hauling a yellow bulldozer behind a big diesel truck. The contractor leaned out of his high window and called "What do you see?"

One of the men at the edge of the pond turned. "It looks like a swan."

"A swan? Really? A wild one?" The contractor set his brakes, opened the door of his truck, and jumped to the ground. "Where is it? Oh, there, over there. Gosh, I've got to call my wife. She loves birds!"

Within minutes the traffic was lined up bumper to bumper for a half a mile, along the edge of the pond. No one raced an engine or blew a horn. Instead, everyone got out to peer across at the bird. Not since a mother bear and her cubs had invaded a beekeeper's yard on the other side of town had wildlife received so much notice. On that occasion, everyone had kept his distance while the bear knocked over a beehive and helped her cubs scoop out the honey, despite the bees and the beekeeper.

The naturalist telephoned news of the swan to a professor at the nearest university. The professor grabbed his spotting telescope, gathered as many students as could squeeze into his little Volkswagen, and raced to the pond. He had to park a long ways away, because so many cars had stopped ahead of him. But he carried his telescope up to the pond and set it up on its tripod. After he and his students had watched the swan a while, he let others see it magnified too. The powerful lenses made the bird seem only a few feet away.

Two more students arrived on a motor scooter. After a quick look through the spotting scope, they took off again. Circling the pond, they rode into the field belonging to the selectman, intending to drive right up to the place where the swan had found a refuge. But the selectman's housekeeper and the naturalist were between the students and the bird. The naturalist talked the students into using his bird glasses to see the swan, instead of going closer and disturbing it.

By noon on Monday, half of the people in the town had seen the swan, and the other half were on their way to take a look. As if it wanted to give everyone a good view, the bird swam out to a shallow place near the middle of the pond and finished preening. A visitor from outer space could not have attracted more attention. Everyone was talking about the swan.

The magistrate of the little town lived in a house overlooking the pond. Through his picture window he could see the white bird wherever it swam. He sat by the hour in his warm house watching the swan and the people who came to look at it.

In the afternoon, he met the naturalist's wife as she was going to shop for groceries. "Are you buying some food for the swan?" he asked her.

She really had not intended to, for the swan had so much wild vegetation to eat—roots and fruits, leaves and seeds—in the shallow parts of the pond that there seemed to be no need.

"Here's a dollar," the magistrate said, holding out a bill. "Buy something for the swan and feed it for me. I've never seen anything so beautiful on the pond before, and I'd like it to stay."

"But if I buy bread for the swan, how can I send it across the water to the bird?"

"Throw the bread on the water. It'll float, and the swan'll find it."

So she examined all the different kinds of bread in the store, and chose the largest loaf with the whitest bread. She hoped it would float as the magistrate suggested, and that the swan would see it. Enough money remained to buy more loaves later in the week.

Just before nightfall, the swan set out to explore the pond. On her journey she found some of the bread that had been torn apart and set afloat. The pieces had become soggy and had settled among the water plants in the shallows. She gobbled the bread, and went to sleep with a full stomach.

The magistrate went to bed with a full stomach too. He dreamed all night of the white swan sailing back and forth across the pond.

The Whole Town Is Swan-Conscious

THE SUN ROSE red over the estuary as the tide flooded in. It glowed on the white swan, who was bathing herself elaborately in the middle of the pond. She dipped one shoulder at a time under water and sluiced as much as a basin would hold over her back. Her neck served as a scrubbing brush. It was so long that she could reach with it all the way to her upturned tail. Her injured leg seemed a little better. She still swam with her left foot, but she let the right one trail into the water. With the webbed toes she could make little paddling movements.

Herring gulls circled the millpond and noticed the swan. Two of them wheeled, calling harshly to one another, and alighted on the dam. The swan moved away from them into the gentle current produced by the inflowing river. She headed upstream, where the woodlands grew close to the shore.

Other birds too enjoyed the seclusion of the woods. Nuthatches defied gravity on the tree trunks and branches, hopping down or up in every direction, seeking insects in crevices of the bark. Their voices, as nasal as a New Englander's, identified them and noted their position: "Yank-yank, yank-yank," they called. Somewhere near, a downy woodpecker hammered on a dead limb with a sound like a typewriter. White-throated sparrows, who were scratching for food among the fallen leaves, whistled clearly at intervals in a pattern of notes whose rhythm matched the words "Old Sam Peabodeeee."

Bright fallen leaves, drifting silently on the current, passed the swimming swan. Slowly the swan traveled up the dark waters. Holding her head high, she watched alertly as she explored the unfamiliar channel.

From a house far above the river, a dog came racing toward the water, chasing a gray squirrel. The squirrel dropped the acorn it was carrying between its teeth and darted for a tall tree at the river edge.

Seeing its quarry leap out of reach, the dog barked loudly and danced around the tree. The sudden commotion startled the swan. The big bird whirled and hurried back toward the pond where she felt safe.

By mid-morning, when the high tide filled the estuary to its fullest, the swan had reached the edge of the dam. There she could look over and see still more water. From the bottom of the dam, a shining waterway spread under the highway bridge, all the way to the ocean a dozen miles to the east. This was the direction from which the swan had come originally. Arching her neck to look down the dozen feet to the tidewater below, she examined particularly the old metal gates that once had opened to draw water power for the grist mill, before the mill tumbled down and the miller changed to steadier employment.

Perhaps the swan would have stayed there longer. But two drivers on the highway slowed to a stop on the bridge close to the dam, to look at the big white bird. An oil truck pulled up behind them, and the oil man stared through his window at the swan. Another car halted behind the oil truck. An intercity bus decelerated behind the car, with a great snuffing of air brakes. The bus driver was behind schedule, and he couldn't see the swan. Impatient, he pressed the

Highland Park Public Library

button on his loudest klaxon horn, demanding a right-of-way. People in the cars ahead responded by gunning their motors to get started. The commotion startled the swan, and she turned back to the open water of the pond. Traffic flowed normally again along the highway.

The townsfolk came again and again to park their cars beside the pond. In the shopping district and all over the community, people no longer said "Good morning!" or asked politely "How are you?" Instead, they inquired, "How is the swan?"

Across town, the swan became a part of the local school activities. The kindergarten children cut out white paper swans to decorate their classroom windows. The English teacher reminded one class that the Ugly Duckling in Hans Christian Andersen's story grew up to be a beautiful swan, and assigned another to write a composition on swans in literature. The music teacher talked about swans in music. She told her classes that the birds had inspired the Russian composer Peter Tschaikovsky to write some of his loveliest ballet music. She played a bit of the lilting and graceful "Swan Lake." Then she showed how differently the German composer Richard Wagner had used an idea from swans. She put on a record of his "Ride of the Valkyries," in which the sound of a swan flying is imitated to suggest great white wings beating the air. All the children heard it in the background—oompah, oompah, oompah. It helped them imagine the awful and beautiful maidens with wings, the Valkyries, who escorted the heroes killed in battle to their mythical home in Valhalla.

The teacher of biology at the high school set aside the laboratory work he had planned, and instead told his students about the swans of the world. He pointed out that swans are the largest waterfowl in the world and the heaviest of all birds that fly. All swans of the Northern Hemisphere have white feathers. The ones with the broadest wing span are the loud-voiced trumpeter swans of Montana and Wyoming. Some of them have a spread of as much as eleven feet. Recently these birds have been given protection from hunters, since only a small number of them remain in existence. They have made a good recovery from the edge of extinction.

The noisiest are the whooping swans of the Arctic in Europe, which nest in Iceland, Finland, and adjacent Russia. The quietest of all are the mute swans of Europe, the kind on the pond. They grow to be the heaviest, to 50.6 pounds, and perhaps the oldest, to 102 years. In between are smaller birds with medium voices: the Bewick's swans, which live in northern Russia, and the very similar whistling swans of the Far North in Canada.

There are also swans with black feathers. Black ones with only a few white feathers on their wings are native to Australia, where they are the national birds. Black-necked swans, which have white feathers except on their heads and necks, live in southern South America.

The physics teacher got out his maps of the stars in the sky and showed his pupils the pattern they could identify after dark as the constellation Cygnus, the swan. It received its name centuries ago and is told about in the Greek myths. According to the old story, a beautiful girl was changed into a swan and placed in the sky, to be admired all over the world. The stars are supposed to mark the shape of a swan in flight, with its long neck outstretched and its wings spread, forming a cross. The brightest star is called Deneb, and at midnight in July, it is almost overhead in the Northern Hemisphere. But in October, the stars of the swan are far toward the northwestern part of the sky. They sink below the horizon before morning.

Other groups took advantage of the swan's presence. The mothers of the Brownie Scouts decided to take their girls down to see the bird and get them out into the warm October sun. When the last school bell rang, the girls came running into the sunshine. The Brownie mothers scooped them up into station wagons and smaller cars. Down to the Community Church they drove, to park in the paved lot overlooking the millpond. In single file the Brownies followed their leaders to the water's edge.

One mother hurried ahead to ask permission to take the troop over private property to one end of the dam. The swan had returned to that vantage point. Perhaps it had noticed that the tide was out. Below the dam and under the highway bridge, only a narrow stream threaded between the big boulders that had been covered completely when the tide was full.

For some reason the swan was attracted to the excited chattering of the Brownie Scouts. Instead of swimming away, she came closer and looked at them. The children stood shoulder to shoulder, the shorter ones in the front row so that all could see. The swan arched her graceful neck, raised her wings just a little, looked this way and that, and wagged her tail. Back and forth she sailed, less than twenty feet from shore, letting everyone admire her. She seemed fascinated by all the different little sounds that came from her delighted audience.

Soon the Brownie mothers grew concerned that someone would fall into the millpond—or perhaps get shoved in accidentally. So they lined up the children and marched them off for a nature walk through an open field, where milkweed pods were opening and the goldenrod heads were dry.

Two small boys, who had joined the throng at the pond edge, remained at the shore. Without thinking, one of the boys picked up a short stick and tossed it toward the swan. As though affronted, the bird turned and sailed away toward the center of the pond.

Almost everyone in town found an excuse to stop for a few minutes at the edge of the millpond to catch up on the latest news of the swan. The milk truck stayed a while. The chief of police parked his prowl car, its motor running, its two-way radio jabbering away, but its blue revolving light turned off. The priest from the Roman Catholic church chatted with some of his parishioners along the shore. The carpet cleaner brought a snack to eat while he watched the swan from his delivery van.

Again and again people asked, "How is the swan's leg?" or "Will the swan stay?" Everyone felt proud that the swan had chosen to settle in their millpond.

For a long time a leader among the town's clubwomen stood by the pond, talking to people. Someone commented that the swan should have a mate. That idea started a whole train of action in the clubwoman's mind. "Let's appoint a Swan Committee!" she said. "The Committee can arrange to get another swan to keep the wild one company. We can have an auction or a public lecture for the benefit of the swan. If we start now, while the swan's leg is healing, perhaps we can induce the bird to stay."

"But is the swan a male or a female? A cob or a pen?"

"The naturalist says it's a young bird. He doesn't know."

"How do you tell the sex of a swan? It would be awful if the town got another bird of the same sex!"

"We can't even give the swan a name until we know what sex it is!"

"All right," the clubwoman said, "the first task for the Swan Committee will be to find out the sex of our bird."

Swan Story

A FEW YARDS from the shore, the young swan floated regally on the millpond. Already she felt accustomed to her new home. With little effort she found food whenever she was hungry, and quiet places in which to spend the night. She had discovered no other large birds for company. But she could not feel lonely, for people kept visiting and bringing slices of delicious bread to her to eat.

Preening took a lot of her time each day. Her second set of white feathers was still growing, although most of the first set had come out several weeks before. The new winter feathers were thick and warm, particularly over her back. They would keep her comfortable no matter how cold the weather turned.

She looked much different now than she had not too long a time before. When first she pecked her way out of her egg shell, she had weighed only half a pound. Her beak had been dull black, like her feet. After a day in the warm summer air, her down feathers had dried to give her a fluffy soft covering. All of them were pale gray.

Her parents had led her to the water, along with her brothers and sisters. At first the older swans had pulled up green plants for her to eat while she floated on the surface. After about two weeks she learned to tip her body and use her own long neck to reach for these flavorful foods. Often her parents let her ride pick-a-back, snuggled into the thick feathers between their wings. From there she could raise her head and look around in all directions. It was quite a contrast from inside the egg shell, which was just four and a half inches from end to end.

By midsummer she had been much bigger. She had had more downy feathers, all of them soft. People called her a cygnet, as they did all young swans.

Toward the end of her first summer, long gray feathers grew on her wings. Her parents encouraged her to fly. At first they flew ahead of her short distances across the water. She could not keep up, although she tried. As soon as she and the other cygnets lagged behind, her parents splashed down to let them all catch up and rest a while. Before winter ice spread across the ponds, they had to be able to fly in search of open water, where they could reach green plants beneath the surface. By then she weighed about sixteen pounds.

The following spring, now eighteen months ago, she had had a new coat of feathers that were a medium brown. Her beak changed color, most of it to a grayish pink, but with shining black at the very tip and around her oval nostrils, and next to her head. As though this was tremendously important, her parents ceased to recognize her. Instead of letting her stay close to them as she had before, they kept her at a distance. The father bird chased her away from the old nest, which the mother bird was rebuilding. He kept bringing more materials for her to use: bullrushes, cattails, reeds, and cord grasses. As fast as he piled them close, the mother bird tucked them into the nest all around her.

The cygnet, then almost a year old, never got to see the six big eggs laid in the nest. She watched both parents take turns at keeping those eggs warm. For forty days she studied them from a distance as they waited for the eggs to hatch. The father bird protected the nest fiercely, although he walked awkwardly on land. Finally he escorted his mate and five baby cygnets safely to the water. The remaining egg never did produce a baby swan. After the swan family ceased to defend it, a skunk came along one night and ate the egg that did not hatch.

Once the year-old cygnet climbed ashore to examine the empty nest. Several times she swam over to look more closely at her parents' new family. But her own father rushed at her, spreading his great wings and hissing loudly. He behaved in this same aggressive way whenever a boatful of people approached. Even the mother bird hissed at her. They were still defending the territory around their nest, driving off any swan or goose from the whole area.

The young swan had learned to live with other young swans and by herself. Her parents had already taught her all they could. And from them she had inherited a large number of automatic reactions that were important to mute swans. Like a library of instruction books, they were filed away in her brain, ready for use when the situation called for them.

Some day when she was older, she would meet a male swan of her own age and kind, one who still had no mate. If, very early in the spring, the two of them discovered an unoccupied territory suitable as a place in which to raise a family, each of them would know exactly what to do. They would begin swimming close to one another, turning right together and left together. In unison they would make the exaggerated signals of a swan's nuptial display. They might swim with their necks straight as rulers, their heads bent sharply downward until their beaks hung vertically. Suddenly, both might hold their heads and beaks level and excitedly jerk the tips sharply upward toward the sky and down again. They would turn to face each other as though dancing on the water, dipping their heads, even immersing their beaks or most of their necks.

Ordinarily, they would go through this elaborate courtship only once. These complex reactions to one another are needed to insure the full cooperation of the two birds in building their first nest. For all later nesting they have experience to follow. And experience is richer than instinct.

All of the courtship gestures could wait until the swan was five years old. Now, in her third year, she was not yet fully grown, although she wore the pure white covering of a mature bird. She remained unaware of whether she was a female bird or a male. She was not yet ready to accept a mate who would accompany her for life.

The naturalist was equally uncertain of the sex of the swan on the pond. He got as close as he could to the bird, and sat quietly studying her from the edge of the water between some dogwood shrubs. He kept looking at the knob where the beak joined the forehead. That knob had a long way to grow before it reached the size characteristic of a mature female. If the bird was a male, it would grow still bigger. Without another swan of the identical age for comparison, no one could judge whether the knob was small or large for the age. Nor could the thickness of the neck tell him anything, although the neck of an adult male swan is thicker than that of a female. The male also holds his neck straighter, with less of an artistic curve. A gamekeeper with long experience with swans might be a good judge. But where could the naturalist find an experienced gamekeeper to look at this single swan?

For a minute he let his attention wander from the swan to enjoy the musical song of a tree cricket. Suddenly he saw the pale insect, for it contrasted in color with the dogwood leaf it stood on, already purpled by the tints of autumn. The cricket was a male, for only they raise their bubble-thin wings and fret them together. They call out a claim to territory as though advertising for suitable mates. The cricket continued his call minute after minute, as a tremolo on a single note. At the same time the insect listened with the delicate oval ears in his forelegs for any competing songs of other males of his particular kind. While his wings were raised in the singing position, he released a perfume as a lure for a mate. If a female approached, her forewings would be slender and pointed, not flat and broad like his. With them she could make no sound at all.

The swan paid no attention to the tree cricket. Nor did the big white bird show any interest when a belted kingfisher flew over the pond. The kingfisher stopped to hover in midair, then plummeted into the water. In a moment it broke into the sunshine again with a small fish in its beak, and swallowed the fish while flying to rest on a dead branch of the overhanging willow. Before reaching its perch, the kingfisher sent out a rattling cry, as if it were exulting over the successful dive. Neither the kingfisher nor its sound had importance for the swan. But the naturalist noticed a brown band across the kingfisher's breast, below the blue belt worn by both sexes. The brown band marked the bird as a female.

About a dozen evening grosbeaks flew into a red cedar tree close to the edge of the pond. Immediately they began chattering loudly, disputing with one another over the soft edible coverings on the seeds of the cedar. The grosbeaks had arrived for the winter earlier than usual, and the naturalist wondered if this meant that cold weather would soon follow. He noticed that some of the grosbeaks wore the golden markings of the adult male birds, and assumed that the rest of the flock were females and young of the year.

His attention returned to the swan. Over and over he asked himself how he could learn its sex without catching it and examining it closely. And he wondered if it was old enough for him to tell even then.

The postman paused to look at the swan. His pouch contained only a few letters to be mailed upon his return to the post office. The naturalist stood up, stretched, and walked into town with the postman. They were good friends, with a common interest in the out-of-doors. Besides, the naturalist wanted to talk to the game warden, whose headquarters were in the post office. Perhaps someone had reported a missing swan. The game warden knew a lot about birds and might have some good ideas about the newcomer on the pond.

During lunch at home, the front doorbell rang. The club lady was there. "I can't stay a minute," she said. "But I want you to know that several people have each pledged five dollars to the Swan Committee toward the cost of getting a mate for the swan. Isn't it wonderful? Everybody's so excited about the project!" And she left.

The back doorbell rang. Three children stood waiting. The youngest, a boy, was puffing from having run to keep up with the two girls, who were older. The smaller girl kept brushing back her hair on one side, where her golden pigtail had come undone. The older girl looked sad, her round face longer than usual. To the naturalist's wife they all cried out together, "The swan's gone!"

"Are you sure?"

Again they all talked at once. Yes, they'd looked everywhere for it. All around the pond. Even up the river. It was gone! Would it come back? Was its leg better? Had anybody taken it?

The telephone rang. The chief of police was on the line. "The swan has gone."

"Are you sure?"

"I saw it go. It started about the middle of the pond, flapping its wings. It kicked at the water too. It sailed high over the dam, over the highway, and down the estuary. I'm afraid we've lost our bird, but I knew you'd want to know."

The Swan's Gone!

WITHOUT THE SWAN, the pond seemed empty and ordinary. Two little grebes came up for air, looked around, and dove again to hunt for food upon the bottom. The kingfishers were patrolling the upper stretches of the river, so no sound of their rattling calls reached the pond. A spotted sandpiper soared to a stop on the lip of the dam, teetered there briefly, and flew off down the estuary. A pair of bluejays began jeering "Thief! Thief!" at a house cat they spied in the undergrowth. The male tree cricket moved to another leaf on the dogwood and continued his shrilling, his waiting for a mate to come. In the light breeze that had sprung up, his calls could scarcely be heard above the rustle of dry leaves, and his perfume lure blew away too quickly to be effective.

The naturalist sat on a mossy rock overlooking the pond and thought about the swan. He was glad the leg had improved enough for the bird to take off from the water. Yet everyone would miss it. He supposed he'd have to help eat up the second big loaf of white bread his wife had bought for the swan. He much preferred his wife's whole wheat bread, even if the swan took the softer white more readily. Now, suddenly, the sex of the swan did not matter any more. It was a good thing the Swan Committee had placed no order for a mate! Almost everyone in town seemed ready to contribute to the Committee's fund. Money seemed to be no problem.

Come to think of it, why wasn't that the answer? Let the Swan Committee buy two swans—a pair, already pinioned so they couldn't fly away, to decorate the town's pond all year. Why hadn't someone thought of that before? The naturalist hurried home to make some telephone calls.

As he entered the kitchen door, his wife was talking to someone on the 'phone. "Here he is now," he heard her say. She held her hand over the mouthpiece. "It's the English teacher from the high school.

You know she lives down on the estuary, next door to that retired plumber. The plumber's wife has seen the swan today, swimming among the sailboats moored in the estuary."

Eagerly the naturalist took the telephone. "Hello!" he cried. "That's wonderful news."

He got no answer. He tried again.

"There's nobody on the line," he said, turning to his wife.

"You might as well hang up, then. You've been cut off."

A few minutes later the 'phone rang. He picked up the receiver, and immediately a voice greeted him. "The plumber's wife called me to the door, and I hung up absent-mindedly. Please excuse me! She says the swan's gone—nowhere in sight. She's sure it's not among the boats any more. I guess we've lost it again. I'll call you back if it reappears."

For some time the naturalist tried to talk his wife into driving with him in their car down along the estuary to look for the swan.

"What good would it do?" his wife kept asking. "If the swan isn't in the pond, the town has no swan. If it's in the estuary, it's on its way back to wherever it came from."

He couldn't argue around her logic. But he did tell her his idea about buying a pair of pinioned swans with the money collected by the Swan Committee. She wasn't very enthusiastic about it.

The back doorbell rang, and he opened the door. The same three children stood there, dancing up and down. The little boy was puffing hard from running. The younger girl's pigtail was coming apart again. But the older girl's face was rounder and happier than usual. As though they had rehearsed the chant, they all sang out at once: "The swan's back! The swan's back! Mummy says to tell you the swan's back!"

"How do you know?"

"We just came by in the car. It's there—on the pond!"

The Royal Bird

WHEREVER SHE chose to be, the swan seemed very special. She had an air of dignity that distinguished her completely from wild waterfowl. Yet she could scarcely be regarded as tame, although she showed no real fear. She belonged to no one. Her comings and goings proved her independence. She acted more like a New England citizen than a feathered creature.

Although it could fly again, the swan remained. Its continued presence was like a testimonial that the town was an ideal place to live. Anxiously, the people sought ways to encourage the bird to stay. They talked about "keeping" the swan, as if they had discovered it and could claim it as their own.

The naturalist knew that swans of this kind enjoy a position of unique distinction among the birds of Britain. There the keeping of swans is subject to an intricate system of customs, laws and ordinances. To learn whatever he could from these charming traditions, he sent off several letters of inquiry by airmail, and waited for replies as patiently as he could.

In England, mute swans have been royal birds ever since 1482. In that year, King Edward IV decreed that all of them belonged to the Crown. "No person of what Estate, Degree, or Condition he be (other than the son of our Sovereign Lord the King)" shall be permitted to keep a swan of his own "except he have Lands and Tenements of the estate of Freehold to the yearly value of Five Marks above all yearly charges." Twelve years later, King Henry VII established the punishment for taking a swan's egg at imprisonment for a year and a day, with a fine in addition if the king so elected.

Each summer, during the week that includes July 15 (St. Swithen's Day), a census is taken of the swans on the Thames River between London and Henley. For centuries, the privilege of owning swans on the Thames has been granted by the Crown to just two of the ancient Livery Companies of the City of London. These companies have become charitable associations of distinguished citizens. According to tradition, their privilege carries with it the responsibility of counting the breeding pairs and the cygnets of the year. This event, conducted with impressive pageantry, is called the Swan Upping.

Beginning on a morning chosen in the Royal Order for Swans, the officials assemble. Successive parts of the river are entrusted to Deputy Swan-Masters, who appoint Rangers to help in overseeing the Rowers and the Waders. Rowers in small boats and Waders in hip boots are hired for the day. The men catch up the swans and cygnets in skillful fashion, capturing them with special swan-hooks. These are thin poles twelve to fifteen feet long, ending in a hook like a shepherd's crook. Only an official swanner may use such a hook, for great care must be taken to keep each family of swans together while the ownership of every bird is settled.

Each new cygnet is given a distinctive swan-mark to show whether it belongs to the Crown or to one of the Livery Companies. A swan-mark is a lasting pattern of shallow scratches and notches in the beak, the scales on the legs, and the webbing of the feet. Once a bird has been marked, its ownership can be determined at any time.

If both parent swans with a brood of cygnets show the swan-mark of the Crown, all of the cygnets are given the same mark and belong to Her Majesty, the Queen. If both parents show the registered swan-mark of one of the Livery Companies, all of their cygnets will be marked identically and belong to the Livery Company. But if one parent belongs to the Crown and the other to a Livery Company, the Deputy Swan-master must divide the cygnets between the two owners. Often the number of cygnets is uneven, and he must decide by lot which of the owners is to be assigned the odd bird. Carefully he records every transaction, as well as the location and size of each brood of swans, in his official Upping Book.

Sometimes the cygnets hide among the reeds along the river and are missed by the swanners during the Upping. These birds remain unmarked, only to be caught the next year when they are bigger. All swans found unmarked belong to the Crown, and are given the royal swan-mark as soon as they are discovered.

The Clerk of the worshipful Vintners' Company, one of the Livery Companies, which received its royal charter of incorporation in the year 1365, wrote to the naturalist:

"that, by custom from time immemorial in England, which amounts to Common Law and is thus enforceable in the Courts, all white swans which are not marked and which have gained their natural liberty and are swimming on an open or common river may be seized to the Sovereign's use by the Royal prerogative."

He gave the census figures from the Upping along the Thames River in 1965 as 1,207 birds. Of these, 185 now bear the Vintners' swan-mark, 161 the mark of the other respected Livery Company (the Dyers' Company), and 861 belong to the Queen. This shows that the swans are thriving, for there were only 559 of them in 1941, and 330 of them in 1884.

Despite the barges and steamboats on the Thames, about 490 families of swans find a place to nest along some twenty-five miles of river. That is nearly twenty nests to the mile.

The letter from the Vintner's Company and another from H. M. Swan Keeper, Rivergate, Cookham Village, Near Maidenhead, Berkshire, convinced the naturalist that there was plenty of space on the mill-pond for a pair of swans and their young. But neither gave any help in determining the sex of the bird.

The Swan Becomes Famous

THE TOWN SWAN made the front page of the Sunday newspaper. A handsome photograph accompanied the description. It stirred readers from miles away to come and visit. Everyone had questions, or information to offer.

In the crowd at the pond watching the swan, a man in a plaid zipper jacket spread the exciting news that he had seen *two* swans on Wednesday, both below the dam. When questioned closely, he conferred with his wife in the car. She corrected him: one swan above the dam, one swan below the dam. Everyone quickly realized that it was the same bird, first in the one place, then in the other.

A middle-aged lady who lived in a white house near the sharp bend far up the river reported that she had seen the swan swimming past her property. It sailed along as smoothly as if it had been drawn by a string, perfectly white against the dark reflections of trees on the opposite shore. She felt that the presence of the swan increased the value of her land.

A young couple on bicycles stopped to point out the swan to the baby they had in a luggage basket hung on the man's handlebars. They had also seen the swan the night before as they came home from a party around midnight. It was cruising across the pond like a ghostly ship in the moonlight.

An oyster fisherman, who was still wearing his high rubber boots, overheard them. He remarked that he didn't doubt they had seen the swan by starlight, but there was no moon, for it was a mere sliver in the sky, rising just before dawn.

An old man who had a black springer spaniel on a leash said that it was probably his fault that the swan had been swimming at night. He had driven home just before midnight and, for a moment, had turned his car to shine the headlights on the swan. The bird had been sleeping on the low marshy island near the middle of the pond, its head tucked into the feathers

New arrival on the millpond.

The swan and other local residents enjoy the millpond.

The swan explores the stream above the millpond.

A close-up view.

Alice in late fall.

Alice in winter quarters with the friend who fed her.

Alice and the professor.

Return to the millpond
—April 1.

OTHER SWANS

South American black-neck swans—and a cygnet.

OTHER SWANS
Australian black swans and cygnets.

Alice and Hamilton—first encounter.

Alice and Hamilton in the backwater

on its back or under a wing. The light had awakened the swan, for it raised its head and turned away. He was sorry he had disturbed its sleep.

A member of the fire department suggested a way to catch the swan when the pond began to freeze. He was sure the job could be done without exciting the bird or hurting it in any way. He claimed that he had picked up in his bare hands more than a dozen wild mallard ducks that had eaten some cracked corn soaked in whisky. For a day or two after, the ducks could not walk straight, and during that time they had gotten used to living behind a tight fence. He believed the swan could be made equally intoxicated and gentle, with no harmful aftereffects except a headache.

One of the students from the nearby university offered a better way to capture the bird. He said that the previous February he had put on his skin-diving suit to catch a mute swan swimming in the estuary near Salmon Falls. Protected by his rubber suit and his aqualung, he had come up underneath the bird and grabbed it by the feet. It was no trick at all. He admitted, however, that the swan was feeble. The bird had died the next week; it was found that its body was full of lead shot.

A gray-haired farmer with a muddy pickup truck felt he had a still simpler way. He used it all the time with his flock of geese. He got someone on the shore near the birds to hold one end of a long rope while he took the other end in a boat and rowed around his geese. So long as the rope was kept about a foot above the water, it formed a barrier. With it he was able to nudge his geese onto land. The same could be done with the swan. Since it could not fly from land, it could easily be herded into a swan house for the winter. The main thing, he stated, was to get the swan house ready for the town's bird.

By the time the naturalist arrived, the conversation had moved on to spring and a mated pair of swans. He was asked about nesting sites for swans on the millpond. He replied that any of the little islands would be suitable. For a new nest, the mother bird gathers twigs, sticks, dead leaves, and soft plant materials, and some feathers from her own breast. She pulls them around her and under her, using her own body as a form. She lays from four to eight eggs, one every other day, each egg with a shell that first is bluish green but soon turns to an olive brown. She makes no attempt to conceal her nest and never seems satisfied with it, continually rearranging the pieces, adding sticks and plant fibers, and enlarging it to as much as six feet across. By the time the eggs begin to hatch, she may have thickened the nest under them until they are two feet above the ground. With luck, all of this might be watched from the road by the millpond.

He warned, however, that in America it is hard for mute swans to raise their cygnets. Particularly in the eastern United States, rain and cool weather are common in the months when the young swans are most vulnerable to getting wet and chilled. In the European countries where swans breed, the weather is better. Moreover, European swans are never attacked by snapping turtles, for these big meat-eating turtles live only in America.

"But first," the naturalist decided, "we must be sure our swan wants to stay. Then we can find out what sex it is, help it through the winter, get it a mate, and see where it will build a nest."

Swan Talk

THE SWAN awoke early. Before sunrise it glided across the millpond. Bending only its feet at the ankles, it seemed to move without doing anything at all. It was using both feet equally now, sweeping first one, then the other through the water shadowed by its floating body. Soon it was far down toward the old dam.

It could hardly have been farther from the point of land where the naturalist and his wife brought food for the swan each morning. Yet it saw its friends arrive and turned immediately to swim straight to them. Behind the bird, the sun brightened the tops of symmetrical V-shaped ripples, which spread toward the shores.

The swan slowed to a stop just before it reached the water's edge. It sloped its neck forward and tipped up its head until the whole underside of its beak was visible.

"Runh-nh-nh-nh," the swan called, as though in greeting. Then it nodded twice and made soft little grunting sounds in its throat, "Huh-huh, huh-huh." Eagerly it accepted some lettuce leaves, first washing each one by swishing it under water, next raising its head to munch a little, finally swallowing.

As though to get closer to the food, the swan swam right to shore. Then it leaned to one side, shuffling a big black foot under its body. Leaning to the other side, it adjusted the second foot. Leveling its body again, the swan arose on both legs and slowly waddled up the slope of land.

At that moment a red squirrel churred somewhere high in a pine tree. The swan raised its head as high as possible and looked around in every direction. It seemed ready to receive a message, a communication of some kind from the squirrel about what was happening that could be important. Seeing neither the squirrel nor whatever had caused the alarm, the bird resumed its feeding.

After the swan finished its breakfast, it stood right

there beside its two friends and began to arrange the hundreds of feathers on its supple neck by pressing the edge of its beak against them. Every few seconds, it paused to adjust an individual feather or to brush out one that had come loose. For a long time it worked its beak through the stiff long feathers of its tail, making sure that they fitted together properly. Sometimes it took a feather into its mouth and expertly hooked together the places where the vane had separated, until the feather was as good as new again.

Shifting its weight to stand on its left leg only, the swan stretched its right leg and right wing to the side, as though to get rid of stiffness from long standing overnight. It stood on the right leg and did the same with the left leg and wing.

Raising a small group of feathers above the base of its tail, the bird exposed a lemon-yellow gland. Reaching around with its beak, the swan rubbed the gland vigorously until oil poured out. It mopped up the oil on the feathers of its cheek and throat. Then, shutting its eyes, the bird used its head to spread the oil over its neck, its underparts, and the "long winter underwear" that came down its legs to the ankles. Time after time it returned for more oil, always nibbling a little to start the flow.

Finally the swan stretched both wings above its back, but did not unfold them. It reached upward with its beak, in line with its neck, and shook itself all over. Settling its wings again, it nodded several times as though getting ready to do something. Instead, it grunted "Huh," and looked straight ahead. Suddenly it called loudly "Who-EEEEE-you," like a musical imitation of an old-style automobile horn. A little cloud of steamy breath from its beak drifted away and vanished in the chilly morning air.

The swan waited, as though listening intently for another of its kind to respond. Or even for an echo from across the water. But none came. The swan called again, "Who-EEEEE-you," and again and again. The three notes rang through the stillness. The only sounds that followed came from a nuthatch, probing for insects in the bark of an elm: "Yank-yank, yank-yank."

The swan bent its neck and pulled off all the remaining fruits from a turtlehead, mouthing them and dropping them. It wasn't hungry any more. It was just lonesome, longing with an inherited urge for the company of another swan.

The big bird waddled down the slope and shoved off into the water, as though setting out to find a mate.

It headed toward the dam, then turned and swam toward the river that flowed into the millpond. As the swan got farther away, the naturalist and his wife rose to their feet from the big flat rock on which they had been sitting. They closed the plastic bag in which they had brought food for the swan and started for home.

Immediately the swan turned and swam directly toward them. "Who-EEEEE-you!" it trumpeted, and arched its wings while sloping its neck forward, its head raised into line. "Runh-nh-nh-nh!" came the strange deep-throated sound, like a prolonged grunt.

"For a mute swan, it has plenty to say," said the naturalist.

"It's certainly trying to tell us something. I wish I could answer it in a way it would understand. It doesn't want us to leave," his wife replied.

"I think it's saying 'Don't go away. You're all I've got!' It wants company and it's signaling to us because we've been feeding it so often it recognizes us as friends."

The naturalist's wife nodded. The swan nodded, too. Then it turned again and swam to a shallow part of the pond and began to search for wild food in the mud at the bottom.

The Pond Freezes

ICE COVERED the millpond from shore to shore. It extended far up the little river, around the bend and almost to the gurgling rapids. Thermometers in the town registered nineteen degrees Fahrenheit. It was the coldest night so far.

On its favorite island, the swan awakened among the marsh plants. Only below its warm body was the earth soft and unfrozen. The bird's breath steamed in the crisp, calm air. The steam disappeared quickly. No cloud marked the wide expanse of blue sky in the early morning.

A few herring gulls spiraled down and alighted on the ice. There seemed to be twice as many of them, because each bird was mirrored in the glassy surface. The gulls argued with one another about where each should stand, and then quieted down as though they had nothing else to do and nowhere to go.

As if to join the gulls, the swan waddled down to the ice and walked out on the flat, slippery surface. Under the bird's weight, the ice creaked but did not break. Slight ridges gave the ice a pattern in the slanting rays of the rising sun. They showed how the gentle wind had shifted and ruffled the water slightly while the ice was forming. Otherwise, the covering over the pond was clear as glass. The swan could see every water plant below the surface, but reach none of them. Everything under the ice was as safe as if it were behind the heavy glass of a store window.

The swan walked over some air bubbles that glistened where they had been trapped as the water froze. It watched a pond turtle scrambling along among the plants below the ice. The turtle paid no attention to the swan. With one foot, the moving turtle jarred a plant on which two snails were creeping and feeding. The snails let go of the plant. They pulled back into their spiral shells, and sank slowly through the cold water toward the bottom. A fish darted out from among the leaves of another water plant and tried to take the snails into its mouth for breakfast. It gave up, because the snails were too large, and swam back to the concealment of the water plant to wait for something else to happen.

The swan could find no breakfast at all. It didn't even have any open water from which to drink. So it set out across the ice toward the little river and the point of land where the naturalist and his wife came with slices of bread every morning. Today they were later than usual. The swan was later too, for it had been in no hurry to walk over the slippery ice. Now hunger urged it over the most direct route.

While still several yards from the point of land, the swan heard its human friends approaching. It recognized their quiet voices and knew that they would bring food. Excitedly, it raised its head high and drew in a breath ready to call. Suddenly the ice crumpled. Splash! Into the cold water went the swan! Sheets of ice more than a quarter of an inch thick tilted up around it like big panes of broken glass. The swan tried to get its black feet out of the water onto the ice, but the edge kept breaking.

The swan finally paused to extend a greeting, then splashed and crashed and smashed its way to get a piece of bread from an outstretched hand. But because it was still a wild bird and not yet fully accustomed to being so close to people, it hesitated a while before coming for another piece. It sipped some water. It thrust its head far down where there might be plants to eat. It nibbled at some twigs and grass along the shore. It looked longingly at the bread, and then found courage to reach for another mouthful.

Remembering that the swan had been too timid to eat when he tried to feed it after sunset the night before, the naturalist decided that the difficulty had been not the swan's shyness but its eyesight. In the gathering darkness it just couldn't see well enough to recognize a friend.

As the swan ate, its hunger diminished. It took longer and longer to come back for more bread. Between times, it swam among the floating bits of ice and gradually opened a larger area of water. One of the gulls saw the food being thrown. Taking to its wings, the gull flew over and circled low down. The swan turned to face the gull and hissed loudly. It followed the hovering gull to one side of the open water, as far as the ice would let it. The gull immediately darted to the opposite side and picked up a bit of the bread floating there. The swan rushed across, hissing and lifting its great white wings. But the gull was too quick, even settling for a moment on the water to grab several more fragments of bread before the heavy swan arrived. In a few minutes the gull cleaned up everything the swan had not yet eaten.

The naturalist's wife dropped a few more slices of bread close to shore, right where she was standing. The gull would not approach so near. With a scream, it flew away. The swan bobbed its head, still facing the departing gull. Then it swam over as if it were no longer afraid of anything, and ate up the bread, finishing its breakfast without delay.

"The weatherman promises sunshine and warmer today," the naturalist said. "I wonder if the swan will take advantage of the open water as soon as the sun melts the ice. It should fly down to the estuary and stay there."

With the Sun to Steer By

AT THEIR MOORINGS in the estuary, the sailboats swung at the turn of the tide and pointed upstream, toward the dam and millpond. The water began once more to flow toward the sea, sliding past the gaily painted hulls. Near one of the boats, a young herring gull floated, paddling just enough to hold its position. It was a two-year-old, no longer brownish gray as it had been its first autumn, but with a dark band on its tail and dark feathers on its neck. It watched unblinkingly as a man rowed out to a small launch moored among the sailboats. Perhaps the man would have some fish to throw away. The gull swam a little closer.

The man got into his launch and tied his skiff astern. He bent his body nearly double to enter the low cabin, and spent several minutes tidying up inside. When he backed out again, he brought with him a long pair of oyster tongs and a handful of seaweed that had dried and shriveled on the tongs since the last time he had used them. He tossed the seaweed overboard, and the young gull flapped over to investigate. A second gull had been flying over high above the estuary. It too saw the seaweed thrown from the boat, and dived with a scream, trying to frighten off the young gull and reach whatever the man had thrown away. The newcomer was a mature bird, and it won the contest. It didn't want the seaweed. Yet it settled on the water and waited to see if something better would come from the launch. A third gull and a fourth, which had seen the second one dive from the sky, came flapping from farther off and joined the gull flotilla around the oysterman's boat. The tide was flowing faster now, and the birds had to paddle vigorously to stay near.

The oysterman started his engine and cast off his mooring. Throwing the clutch into gear, he spun the wheel and headed downstream. He had a long way to go to reach the best oyster beds, and he wanted to be there at least half an hour before the tide ebbed to its fullest low. That would give him the best chance to get his legal limit of oysters for the day.

As his launch curved into the swift current and approached the last few sailboats at their moorings, he advanced the throttle and felt the wooden deck pulsate beneath his feet. He couldn't use full-speed yet, for fear his wash would rock the sailboats more than they could safely stand.

It was only then that he saw the swan, for it had been hidden from view behind the farthest boat. Instantly he cut the throttle back to idling speed and spun the wheel to veer away. But in that same moment the swan saw the onrushing launch. It swirled around, flashed out its great white wings and beat them downward furiously. Kicking at the water with both feet, it helped itself gain forward motion. Soon it was airborne and away. The launch came close. But once clear of the water, the swan speeded up. Low over the estuary it sped off toward the distant sea. In a few minutes it followed the channel around a point, and the oysterman lost sight of it.

The oysterman sighed. If he had chased away the town's swan, he would never hear the last of it. He could only hope the bird would come back.

Down the channel he watched on both sides for the swan, but saw no sign of it. He looked for it again all the way home. No swan.

The wife of the retired plumber saw the terrified swan fly away. She rushed out of the house in panic, in time to see it disappear in the distance. There was nothing to do but spread the news by telephone. Soon everybody knew the bird was gone. To add to the general feeling of gloom, the sky clouded over. A persistent drizzle began to fall.

Telephones were busier than usual because neighbors called one another to talk about the missing swan. Some wondered what it would get to drink down near the sea. They were assured that swans can drink sea water without harm, just as wild geese do. They get rid of the salt through salt glands that empty into the nose. The salt is rinsed away from the beak while the bird has its head under water, reaching for plants to eat.

Only the naturalist was optimistic about the bird's return. "I'm counting on the bird to come back. It's been well-treated. But it won't fly in this direction until it has the sun to steer by. The sun is its skymark. Hidden in each bird brain is a clock of some kind that corrects automatically for the position of the sun: east in the morning, south at noon, and west toward sunset. As soon as we have sun again, the swan will return all by itself."

The sky cleared a little during the night, and brightened in the east before dawn. The wind came from the east too, which seldom indicated any improvement in the weather. But the naturalist got up a few minutes before sunrise. Seeing the broken clouds, he walked quickly down to the millpond. No swan. He continued around, past the dam, to the estuary. The sailboats all faced the sea, partly because of the wind and partly because the tide was coming in.

Perhaps he'd been wrong to predict the return of the swan. Once it got within sight of the sea, it might go up some other river. Or just stay near the ocean. He looked down the channel, but the wind ruffled the water, breaking the reflection of the rising sun and of the bright eastern sky into thousands of sparkling points that blinded him. To the south and west and north the clouds were as thick and dull gray as ever. Probably the sun would soon disappear. Perhaps the swan hadn't had time to come back. Yet he hurried along the shore toward the ramp near the sailboats, where people could launch boats from trailers behind their cars.

"Who-EEEE-you! Who-EEEEE-you!"

He couldn't see the swan against the light. But the bird had spied him and called. Was it recognition? Little tingling sensations ran up and down his spine. Why hadn't he brought some food for the bird?

Standing motionless, he wondered what to do. Should he go to the water's edge and at least talk to the swan, welcoming it empty-handed? Or should he go away as inconspicuously as he could and get some food for the bird? Crouching a little to conceal himself, he moved toward the road and headed home to spread the good news.

Before he could return, the sky clouded over. But he fed the swan near the boat ramp. So did some of the townsfolk, later in the day. They were delighted to see the bird again, but distressed that it made no move to return to the pond.

When the tide reached its high during noonhour and the salty water came right up to the bottom of the dam, the swan waddled out on a muddy bank and just stood there. Some people wondered if they should help it to get back into the water. The lady who sold antiques and puppies talked of calling the conservation officer of the Fish and Game Department. He could come with his truck and a net to catch the swan and take it away to some safe place for the coming winter. Only a few had confidence that the swan would take care of itself. The naturalist kept assuring everyone that when the sun came out, if the tide was reasonably high, the swan would fly back to the open water of the millpond.

Even when the swan was in the millpond, it did not fly when the tide was low in the estuary. Nor did it have to go down to the dam and look over to find out what state the tide was in. It had a mysterious ability to regulate its life according to the rhythm of the tides, an ability shared by smelt and alewife fishes and by many of the crabs that live near salt water. No man has yet discovered how these creatures know so well the rhythm of the tides, for the ebb and flow of coastal water comes about forty minutes later every day than it was the day before. To predict the tidal changes at any place, human calculations require a complete record of the times and levels of high tide and low for several years. Yet without a clock or any other instruments, the swan alters its daily schedule to match the tidal changes. Without any conscious effort, it reacts in ways that improve its chances for survival.

So, while waiting for the tide to be right and the sun to be out, the swan returned to the water without help and plucked bright green sea lettuce from the rocks. It found a safe place above the high-tide level on which to spend the night, its long neck laid smoothly atop its warm back under the snowy feathers of its folded wings. And eventually, it flew to the millpond, just as the naturalist had predicted. It was obviously content with its home.

White Swan in a Snowstorm

THE SWAN PADDLED close to shore among some cattails. In a few places it had to break a thin sheet of transparent ice to go where it wanted. It reached forward with its long neck, sometimes under the ice, to find fleshy roots and edible stems in water more than a foot deep. Often it stretched so far down that the slender part of its neck disappeared altogether. Then its body seemed to end in front as a white cone of feathers at water level.

To get food it fancied from still deeper parts of the pond, it tipped its whole heavy body like an acrobat, pointing its white tail toward the sky. It could reach then straight down with its beak. At other times the swan moved over the place where it had been feeding and used its two feet alternately, treading on the bottom as though rearranging the vegetation there. Its body rocked from side to side, starting waves in both directions. While it rocked, it held its head in the elegant S-curve that artists have admired ever since mute swans began living where people could watch.

Not far from the swan, a high school girl stood motionless, entranced. She had been lucky enough an hour before to see the bird fly over the dam from the estuary and alight in the pond. Over the tops of the tall elms by the highway beyond the dam, the glorious creature had come swiftly. Its outstretched neck lowered a little as it aimed down on a long slope toward the water. It flared its wings wide, like a white statue of winged victory, as it braked to a stop, splashed into the pond, and skidded along a short distance like an amphibious airplane that has just landed on the water. Then it calmly folded its wings over its back, nodded its head a few times, wagged its pointed tail, and swam sedately in her direction as though it had never been anywhere else.

Three more high school girls arrived on bicycles. They had heard that almost anyone could now feed the wild swan. One of them carried a paper bag half-full of stale white bread for the bird. She took the bag with her to the water's edge and began tossing bits of food toward the swan. Without hesitation the swan swam over and began eating. It took each piece separately, thrusting its head into the water almost eye-deep. Nibbling quickly at the bread, it wet the food thoroughly before swallowing. As the bird's beak opened and shut, water went in and out at the sides through a regular row of little projections that served as a strainer. Yet in every movement, the bird maintained its dignity.

While the swan was eating, the girl who had seen it fly back to the pond described her experience to the others. "If it flies down to the estuary or back again," she concluded, "I hope it flies high. It would be awful if it hit the power wires or went under them and was struck by a truck."

"Maybe if we feed the swan all it will eat every day, it will get too fat to fly. Then it'll have to stay in the pond where it's safe."

One of the bicyclists looked up at the sky. "I just felt a drop of rain. Guess I'll get going."

"Who's going to feed the swan during a snow-storm?"

"I'll come even then," promised the girl who had

watched the bird in flight. But she too headed for home before her hair got wet. Already the raindrops were starting little circular wavelets expanding all over the open water of the pond.

In an hour, the rain changed to snow. Soon it hid the ground like a soft fluffy blanket. It hid the tiny gray springtail insects that had been hopping about on the thin ice. It covered the gray stones in the old wall. For the first time that season, the weasel matched his surroundings. Already the sleek little animal had his winter coat, making him ermine white everywhere except the black tip of his tail.

Atop a snowy stone in the wall, the weasel paused and looked over toward the swan. The bird had tucked its head among its feathers and had its neck flat on its back. Standing motionless among the low plants, the swan appeared to be just a big snow-covered boulder. In fact, it *was* covered with snow. So perfectly did the white feathers of the bird hold in the heat of its body that the flakes did not melt. Although in plain sight, the swan had disappeared.

Swan-nappers

THE DAY AFTER Thanksgiving, the weather turned warm. Every trace of the early snowstorm evaporated, leaving the ground slightly wet. The ice melted from the pond, allowing the swan to swim freely again. In the late-November sunlight, she reached hungrily into the water for green plants to eat.

The magistrate felt happier than he had been for a long while. He stood, sipping another cup of coffee, and looked out through his picture window at the swan on the millpond down the hill. It was wonderful to see the big white bird glistening on the water. He set down his empty cup carefully and picked up his bird glasses to focus them on the swan.

Just then two cars came along the road beside the pond. They stopped and three people got out. The swan paid no attention. But the magistrate studied the newcomers through his bird glasses. He didn't recognize any of them. This surprised him, for he knew most of the townsfolk at least by sight. He wondered who the woman was. She wore an expensive-looking coat and a fancy hat. One of the men with her appeared rather distinguished. He had on a well-cut topcoat, a soft felt hat neatly shaped, and ivory-yellow gloves. The other man seemed shorter because he wore a dark cap and a jacket that scarcely reached his hips.

The two men began hauling a rowboat out of the station wagon. They pulled it down to the pond and shoved it part way into the water. Now the swan was watching them almost as intently as the magistrate. The man in the cap went back to the car for some oars. The swan began swimming away from the road, across the pond.

The magistrate called his wife from the kitchen. "What do you suppose is going on down there?" he asked, handing her the bird glasses.

The man with the cap got into the boat with the oars, fitted them into the oarlocks, and pushed off. He began rowing toward the swan. The other man joined the woman at the edge of the pond. She seemed to be directing everything except the bird, which was heading straight for the little island where it could cross into deep water. The man in the boat changed direction, for he saw that he would have to row a long way around the island to follow the swan.

"I'm going down there," said the magistrate. "Nobody can chase our swan like that. Who do they think they are?"

He hurried into his coat and hat and strode down the hill.

"What's going on?" he demanded, when he reached the well-dressed couple beside the pond.

"Our man is trying to catch our swan," the lady explained.

"Who says its your swan?"

"Our man recognized it. In fact, we all did from the photograph of it in the *National Observer.*"

The magistrate introduced himself, and learned that the couple came from the next state, where they kept swans on a small lake in the middle of their large estate. One morning a flock of geese had settled on the lake and, when they took off again, the young swan had gone with them. Their man had never gotten around to clipping its wings. No, it wasn't marked or banded in any way. He hadn't done that either. But the date given in the newspaper for the arrival of the unknown swan on the pond matched perfectly. Besides, the bird out there was exactly the shape and size of the one that flew away. It simply had to be their bird.

The man in the rowboat got stuck on one mud bar after another. The swan kept far ahead of him, outmaneuvering him every time. The man kept glancing over at his employers, hoping that they would give him directions or call off the chase. But they were too busy talking to help him. So he kept on rowing in the direction of the swan. If only he could drive the bird ashore, he was sure he could catch it bare-handed.

The magistrate was trying hard to be fair about the swan. These people were obviously well-to-do and could take care of the bird, whereas no one in town had experience with swans or had offered a winter home for it. He felt sure that the swan would be better off with other birds of its own kind. And yet, if these people took away the bird while the townsfolk found the creature so completely enchanting, there would be a lot of hard words said. He himself would certainly miss the swan, although he realized that it could not stay out on the frozen pond all winter.

The lady decided that she needed more help to catch the swan and beckoned to the hired man in the rowboat to return. When he reached the shore nearest the road, her husband helped the man get the boat into the station wagon.

Since the swan seemed safe for the moment, the magistrate merely said goodbye to the couple. Still wondering what was the best thing to do, he walked home to make some telephone calls.

He consulted the District Attorney.

"No," said the D.A., "it's illegal to transport a wild bird across a state line without a permit."

"But this is a domestic swan," the magistrate objected. "How can you call it a wild bird?"

"If it is unmarked, and you say it is, and it flew in, and you say it did, then it's a wild bird."

"I don't know whether it's marked or not. The people who want to catch it say that the bird they lost was unmarked."

"Then they can't positively identify the bird on your pond. They have no right to catch it and take it away."

"But it's probably their bird," said the magistrate.

"Probably isn't good enough. You can't let them

take it out of state. It's illegal even to possess a wild bird," said the District Attorney.

The magistrate sighed, thanked him, and hung up. He called the police department and told the desk sergeant about the attempt to remove the swan and the ruling from the District Attorney. The sergeant promised to alert the police force and to keep a patrol car near the pond to put a stop to any future attempt to remove the swan.

The magistrate tried to reach the naturalist, but no one was home. Next he called the game warden's office. But the warden was out. He left a message there and telephoned the nearest representative of the State Fish and Game Commission, to learn that an unmarked mute swan would qualify as migratory waterfowl for which no hunting was allowed, no possession permitted. Taking such a bird across a state line would be a serious offense.

Going to his picture window, the magistrate looked down at the swan. It was sailing serenely across the pond as though nothing had happened. "Bird-nappers," said the magistrate to his wife. "That's what those people were. All they wanted to do was a little swan-napping. Well, they can't. But I haven't met nicer people in a long time."

Whose Swan?

The swan was sound asleep in the middle of the millpond, surrounded by a thin film of fresh ice when the naturalist and his wife returned home the Friday night after Thanksgiving. They had not enjoyed the festive season quite as much as usual, for the roast turkey had reminded them too much of the live swan with whom they had made friends. To remember that swans were domesticated long ago as a convenient way to have food available did not help at all.

The telephone rang. The magistrate was on the line, eager to tell about the attempted swan-napping earlier in the day. He wondered whether the Swan Committee had made any definite plans to get the bird off the pond. He did not mention his conversation with the District Attorney, or suggest that a permit would be needed even to house the swan during the winter ahead.

"Who are these people who claim the swan?" the naturalist asked.

"You'll hardly believe me," the magistrate replied. "It sounds more like a fairy tale than the truth. They're a real Prince and Princess, with a big place that's the American equivalent of a castle. The wife—the Princess—is an heiress, and the husband—the Prince—a Russian who escaped at the time of the Revolution. His father ruled some small Russian principality. The wife does most of the talking."

"It seems to me there's one flaw in their story," the naturalist said. "If they recognized the swan, surely the swan recognized them. If their hired man fed the swan for weeks and really knows the bird, why didn't he call it and get it to come to him? It comes to us from all the way across the pond. Either the swan doesn't like these people, or it doesn't recognize them. I don't believe it would forget them so quickly. And only someone unfamiliar with waterfowl would ever go after a swan in a boat, certainly on an area as large as the millpond."

"Then you believe that the town should keep the swan?"

"I'd say instead that the town should provide a winter home for the bird that chose the town pond."

The magistrate seemed reassured and said good-night. The naturalist told his wife about the telephone message. Then he called a professor who had had experience with management of waterfowl on government sanctuaries in Canada. The professor wanted to consult the game warden before recommending any course of action, and promised to try to reach him right away.

Half an hour later, the professor called back. "The game warden knows all about the attempt to capture the swan and take it away," he said. "He insists that unless the bird is banded and the band number corresponds to written records in the possession of the Prince and Princess, they cannot claim the swan."

The naturalist was sure that the swan had no band on either leg. This meant that the bird was technically a wild one and belonged wherever it chose to be. It would be illegal to catch or confine it, and a Federal offense to take it across a state line. The town officers, however, could probably be authorized to provide for its safety during the winter.

The town offices were already closed for the weekend. But the police department remained alert. At 3:15 in the morning on Saturday, two officers in a patrol car discovered the swan walking along the middle of Main Street, fully a quarter of a mile from the pond. After they recovered from their astonishment, one officer guided the big bird slowly down the road while the other followed along in the car, shining the headlights to light the way. For a while the swan tried to turn aside, first into the parking lot behind the Community Church, then into an open field. But the policeman was too quick on his feet and too determined. By 4:00 the bird was back in the water, with instructions to stay there!

About mid-morning a pickup truck with an out-of-state license arrived at the pond and stopped. In it were the son of the Prince and Princess and their hired man. They had a rowboat in the truck and were prepared to catch the swan. By the time they got out of the truck, a police patrolman was parked right behind them, and the game warden had walked over from the sedan in which he had been waiting. The warden advised the newcomers that there would be a $500 fine if they took the bird across the state line without first proving that it was theirs. He explained what kind of proof would be necessary, and indicated that he was ready to compare their record of a number on a leg band with whatever they found on the swan if they caught it. They admitted that they had no such definite proof. Reluctantly, they got into their truck and drove off. The police car followed them to the edge of town.

That evening, the senior selectman telephoned the naturalist. "You'll be glad to know that the swan is now officially a native of the town," he said. "The game warden helped me draw up a legal paper today. I signed it and affixed the seal of the town. It also designates the millpond as a wildfowl sanctuary, and gives the town the right to provide winter quarters and food as necessary for the safety and health of all wild birds on the pond."

As though this were the missing piece in a jigsaw puzzle, a plan of action occurred suddenly to the naturalist. He would get help in capturing the bird as soon as the pond froze over solidly. He would learn its sex by examining it closely, as the first step toward getting it a mate. And then the bird would need some kind of warm winter quarters. He called several friends to share his idea. But when he asked for help in finding a home for the swan, nobody had a good place to offer.

Before he could dial one final number, he was interrupted. The telephone rang. The long-distance operator asked for his wife. He waited, listening carefully, trying to interpret a conversation that ranged over a whole assortment of exotic birds: macaws, parrots, and finally swans.

When his wife hung up, she looked amazed. "The swan's not completely safe yet. That was the secretary of the Princess, calling from out-of-state. Apparently the Prince and Princess still believe they can get a permit to take our bird. They are very pleasant and indirect about it, but they are also determined. The secretary thanks us for taking such good care of 'their' swan. She says they did telephone the town office before coming up the first time, and whoever answered the call seemed quite agreeable. She quotes the town office, 'If it's your swan, come and get it.' They did not expect any difficulty."

"Did she say why they had never clipped or banded their bird?"

"No. But she mentioned that a neighbor had lost a pair of Australian black swans whose feathers had grown in. When the birds were located, there was no objection to bringing them back. I told her that black swans have never gone wild in America. Mute swans are different, for some do live free and protected. Aren't there some near Providence, Rhode Island, and some more down in Portland, Maine?"

The naturalist nodded. "Did she know the sex of the bird that flew away?"

Not for sure. I suggested that if the Prince had a breeding pair and knew the sex of the bird he lost, he would have one supporting fact to offer—providing, of course, our swan proved to be of the sex that was missing."

"If he just guessed, he would have a fifty-fifty chance of being right. I'm sure he'd need more proof than that."

A Home for the Swan

THE LAST DAY of November dawned clear and cold. The pond froze over from shore to shore. Little tufts of ice crystals dotted the surface, as if a chenille bedspread covered the chilly water.

The swan rested on the ice and preened herself. Both of her feet were raised and tucked away among the warm feathers on her back. Inside her own swansdown comforter, the bird was snug. About seven o'clock, she had a little bread for breakfast. The white slices had been skidded across the ice right to her side. But without water in which to wet her food, she wanted less to eat than usual.

Just after nine o'clock, the wife of the naturalist

walked along the road beside the pond. A policeman in his patrol car caught up with her and told her that an hour before he had watched the swan drive off a dozen gulls. They had come to steal the uneaten bread. Without rising, the swan had braced itself with its legs against the ice and spread its great wings, flapping them vigorously. The gulls took no chance of being struck. For them the bread was too close to the swan. Later he saw a black crow settle on the ice at a safe distance and eye the bread. It too flew away. A dog circled the swan, but made no attempt to come close. Obviously the big bird could take care of itself.

The naturalist's wife was on her way to pay their bill for fuel oil. Another customer there spoke of the swan. It was wonderful, he said, the way the bird had come. He wished he had seen the big flock of white swans flying over the town, coming from Hudson Bay in the Far North, going to the Gulf of Mexico to spend the winter in the sunshine. He was glad that this injured bird had gotten tired and settled on the millpond. The story added some new folklore to the history of the town.

"Yes," the naturalist's wife agreed, "it's quite a story! If only it were fact!"

She went to the bank. The tellers stopped counting

money. The girl at the bookkeeping machine turned around and stood up. The people who were waiting to cash their pay checks all asked the same question. When was the Swan Committee going to get that poor bird off the pond and protect it from the weather?

She went to the town office to pay her taxes. "Guess what!" said the girl at the desk. "The secretary of Princess What's-her-name telephoned this morning to say that their swan refuses to eat, now that its mate has flown away. She's afraid it will die if the town doesn't let the Princess have the swan on the pond. I told her to send up her lonesome, hungry swan. We'd look after it."

A housewife next in line at the counter laughed, then turned serious. "Honestly," she asked, "where are we going to put our swan for the winter?"

Like the swan on the frozen pond, everyone was waiting for the the answer. The town didn't even have a jail where the big bird might be given space until spring. Meanwhile, the weather forecaster promised more cold nights and days. Soon it would be impossible to keep a hole open through the ice at the edge of the pond, where the swan could come to drink and eat wet food.

Several people worried about the safety of the bird

resting out there on the ice. Now that dogs could walk across the surface of the pond, a whole pack of them might attack the swan at once.

The Director of the State Fish and Game Commission received a telephone call from the Princess. She pleaded with him to release the swan into her custody. "The town has no place for the bird," she said. "And I can put it in a nice, warm swan house."

The Director called his local game warden. He also got in touch with the representative of the Federal Fish and Wildlife Service. He talked to the magistrate in town, and then to the senior selectman. "Why is everyone so excited about this swan?" he demanded. "I haven't had so many messages about a single bird since I was appointed!"

The naturalist consulted the professor who had had experience with wild geese in Canada. Did he think the swan would find enough to eat all winter if it were merely caught and freed on the open water of the estuary? Dozens of Canada geese sometimes stayed there through the coldest months.

The professor shook his head. He had examined some of those geese in the early spring. "They're thin and weak by March," he said. "They can hardly fly until after the ice melts and they reach better supplies of greenery. I'm afraid the swan would starve to death. At this latitude it needs a place indoors where there is protection from the wind."

"If I find a home for the swan, would you help us catch it?"

"When you're ready, let me catch it," the professor volunteered. "It will hate whoever captures it. Let me be the one it hates. Then it will take food from you and your wife again next year."

So the naturalist continued to inquire about a house for the swan. He had one more idea. He called the head of the animal science department at the university. Wouldn't the department like to demonstrate its friendship for the town by helping find a place for the town swan through the coming winter? The Swan Committee would pay for the food and find plenty of people who would be responsible for feeding the bird and keeping its living space clean.

The head of the animal science department shared the local enthusiasm for the swan. He promised to check with the manager of the poultry farm to see if one of the houses usually used for chickens, ducks or geese was vacant. "It would be unheated, you understand, though we do have electric heaters to put in pans of drinking water to keep them from freezing.

And of course, we could take no chance of introducing any diseases. The swan would have to be tested thoroughly by one of our specialists and be certified free of infection. Even then, if the university provides a place for the swan during the coming winter, I would not want to have this regarded as setting a precedent."

"Oh, no!" said the naturalist, smiling as he listened over the telephone. "Please let me know about the space, though, as soon as you can."

Freedom Interrupted

November went; December came. Nothing in the cold world around the swan or inside its warm body stimulated it to action. It rested motionless atop the ice, its feet tucked in among its feathers. It had been in this one position all night, as if it were waiting for something to happen.

At eight in the morning, the naturalist and his wife appeared, their breath frosting in little clouds every time they breathed. As usual, they brought food in a plastic bag. The swan raised its head and greeted them, but kept its feet hidden. Until it felt hungrier, the distance across the slippery ice was just too great to be worth the effort. The swan watched the naturalist use his boot to break through the thick ice at the edge of the pond, so that the bird could drink some water. It wasn't thirsty enough to make the trip either.

From a point of land beyond and to one side of the swan, a young man in a ski jacket and hip rubber boots skidded out on the ice. He had no confidence that the frozen covering over the pond would support him, for he weighed more than four times as much as the bird. But the ice merely creaked and gave off faint cracking sounds, without breaking. The newcomer was the professor who had volunteered to catch the swan. Slowly, shuffling his feet to keep from slipping, he approached the bird.

Rocking to one side and then the other, the swan got its feet under it and stood up unsteadily. It walked toward the small area of open water where the naturalist and his wife stood waiting. The professor took plenty of time. The bird showed no real hesitation, but it shivered a little with apprehension. It knew that without a strong wind, it could not use its great wings to escape if the people came too close. On such a calm morning, it could not take off from land—let alone from slippery ice.

The professor said something. The naturalist and his wife backed away from the edge of the pond. The swan came to the open water and took a sip. Yet it ignored the bits of green lettuce and the slice of white bread that were floating there. Ordinarily the swan would have enjoyed these foods more than almost anything else. The professor came closer. The swan shivered. Then it stepped into the cold water and floated among the food and chunks of ice.

The professor walked around to the shore. The swan climbed out of the water and stood up on the ice as high as it could. Before it could take more than a few steps away from the shore, the professor was back on the ice, ready to circuit again. The swan got back into the water and began to preen a feather on its back. The feather needed no attention, but it was as if the swan felt it had to do something about the situation and preening and sipping were the only things it could do, even if neither helped at the moment.

The professor tried approaching from a new direction and scattering handfuls of yellow corn ahead of him on the ice. Bending himself nearly double, he got all ready to spring at the swan. Abruptly he straightened his legs and stretched out his arms to seize the bird. Crash went the ice! Splash went the professor! Hip-deep he plunged, clutching only an armful of soggy cattails.

Faster than ever, the swan got onto the ice and waddled away. The professor climbed out too and brushed the water from his clothes. He remained cheerful and calmly started over. Across the pond he went, to drive the swan back to the open water.

The next time he succeeded. He rushed at the bird and almost fell on top of it. Before the swan could raise its wings in self-defense, the professor pushed it deep into the icy water and grasped it around the big strong wing bones close to its body.

The swan cried, "Ahh-h-h!" Its voice seemed almost like the bleat of a lamb, calling for its mother. "Ah-h-h-h!" It was a young bird's voice, quite out of proportion to the swan itself.

The professor got his feet under him securely and lifted the swan out of the water with his hands clutched around its wing bones. He was sure that as long as he held the bird in this way it could not strike him or escape. The swan did not struggle as much as he expected. But it was heavy, as if fat and well fed. It kept stretching its long neck away from him, and pointing its beak toward the gray sky, where it wanted so much to be. "Ah-h-h-h!" it cried.

With the keys to the professor's station wagon in his hand, the naturalist ran along the road to get the vehicle. Swiftly he backed it into position and helped the professor over the lowered tailgate onto a big cloth spread on the floor. Still wearing his hip waders and clasping the swan to him, the professor sat as securely as he could. The naturalist and his wife drove the car as gently as possible to the building where the professor had his laboratory. There the strange procession of one swan and three people went unobserved into a secluded room with a cement floor, where the big bird could be penned safely with plenty of food and water.

The swan had stopped crying as soon as the professor sat down with it in the back of the station wagon. It made no sound until after he lowered it to the cement floor of its temporary pen. Then it stood up, wagged its tail, shook itself all over, and said, "Huh-huh." Only its dignity had been hurt. It walked to one corner of the pen and stood there, facing the angle in the wall.

The professor brought two big plastic trays and set them down in another corner. Into one he poured more than a gallon of clean water. The second he heaped with pieces of lettuce and chopped apple. He had some corn and other grain ready too, but he waited with these until the swan felt more like eating.

Everyone tiptoed away and left the big bird alone, to get used to its new surroundings. They locked the door and put up a sign, "Experiment in Progress. Do Not Disturb. Leave Room Lights On."

The next day, the professor had to hurt the swan's dignity still more. He came with the expert from the animal science department, to test the bird for infectious diseases. Again the professor made a sudden move toward the swan and grasped it around the big bones in its wings where they joined its body. This time he lifted the bird and turned it upside down. Lowering it gently to the floor on its back, he held it securely while the expert took a sample of its blood from one leg. All the time the professor held it, the swan kept crying "Ah-h-h-h!" in its high-pitched voice. When at last he set it right-side-up on the floor, it scrambled to its feet, said "Huh-huh," and walked to its darkest corner, to stand there as though it were being punished.

The expert needed twenty-four hours to complete his tests. The swan, the professor, the naturalist and his wife, simply had to wait. The manager of the poultry farm at the university had to wait too. He was impatient to prepare a winter home for the swan in a vacant poultry house. He planned to tack plastic sheeting on the walls to keep out the wind. He had two electric heaters ready to keep the drinking water from freezing. And he had a shiny padlock with four keys. One would be for the professor, one for the naturalist, and one for whomever the Swan Committee got to care for the swan every day. The fourth key he would keep himself, just in case it was needed.

The health expert completed his work as quickly as he could: the swan was in perfect health. And the professor had also done his job well. He reported to the naturalist: "From my examination, I feel 85 per cent sure that the bird is a female; there is only a 15 per cent chance that it is a very young male."

"Is that good enough?" asked the naturalist's wife, "or will you need to check again in the spring, before we make arrangements for a mate for the swan?"

"I think we can go ahead," said the professor. "Perhaps tomorrow, if all goes well, after we get our bird into her new home on the farm."

The Year of the Swan

FROM MORNING until midafternoon, the sun streamed through the big window into the swan house. The swan looked out and wondered at her world. She could not know that the millpond, where she had been so contented, was now covered by inches of snow as glistening white as her own feathers. The snow whitened the field she could see from the window, and the roofs of the distant buildings too. Winter had come early.

The swan house was cool inside, almost the same temperature as outside. A narrow space protected by wire mesh let in fresh air above a second window. Clean wood shavings littered the floor, keeping the feet of the swan dry and warm. They could be swept out and replaced once a week. Fungus molds were now her greatest danger. Swans, like wild geese and penguins, are especially susceptible to some of the molds that grow commonly on wet surfaces.

Two shallow trays, each twenty inches across, held water. Clipped to each tray was an electric heater with a thermostat to keep ice from forming no matter how cold the weather got. The swan would always be able to get a drink, and to wet her food as she liked to do.

She poked her head into two other trays, separated a few inches from those with water. One contained a fluffy green material she had never met. It was alfalfa meal, full of vitamins and minerals as well as nourishment. The other contained a mixture of corn and smaller grains. The naturalist had dropped a few pieces of green lettuce on top of the corn; it had soon frozen. The swan did not know what to do with it. It took her almost two weeks to learn how to lift the frozen lettuce into the warmed water and make it good to eat.

Two or three times a day she had visitors. For a while she avoided them. She walked over to the corner and faced away while they fixed her food or cleaned her house. Or she tucked her head into the feathers on her back and just watched them with one wary eye.

She grew more confident when a month went by without anyone's attempting to touch her. Finally she even raised her beak high and made contented sounds deep in her throat a few times when the wife of the naturalist came and talked to her. But she could not feel quite the same enthusiasm as she had on the millpond. By winter's end she had made a friend of the kindly gray-haired man who came each day to feed her.

Only her inner clock and the increasing height of the sun in the sky each day told the swan when winter changed to spring. They could not tell her what was in store when, in a jumble of human sounds, the professor who had caught her arrived. He came with the gray-haired man who brought her food. She recognized the professor immediately, and went over to her corner—something she had not had to do for months.

The professor said, "She looks to be in wonderful condition. As soon as the ice goes from the pond, I think we'd better set her free. But to keep her from flying away promptly, I'd like to clip her long flight feathers on one side. She'll grow new feathers in the late spring when she molts. By then she can choose whether to stay or to go."

The gray-haired man nodded. But he looked sad. He did not like to think of clipping any feathers from the swan. Yet the plan did make sense, and the swan would have to be handled to take her to the pond.

"What about the mate you promised to get her?" he asked.

"Oh! Didn't I tell you? We did get one, from the wild flock in Rhode Island. But he was a sick bird. He died a few days after we got him. I'm trying to arrange for another one. It's fantastic how complicated the regulations are for getting a wild swan from another state."

"Well, let me know when you're ready to put our swan on the pond. I want to be there to see how she reacts to being free again."

The professor reached into his brief case. "Did you see this? I picked it up at the town office. It's the report of Town Meeting."

"I was there."

The professor thumbed through to page eleven, and read aloud:

"On the lighter side . . . the Chinese have their new Year of the Horse, while we . . . have . . .

The Year of the Swan. Some people have suggested 'Onderdonk' as a name for the swan, but we would like to suggest calling our rare bird 'Alice' since it seems more in keeping with the town's informality and the general first-name basis of our citizens. Alice's arrival overshadowed almost all other events of public interest. Long may she live."

"I know. I know," said the gray-haired man. "The townspeople appointed a Swan Master, too. But has he ever been over to see our bird? I'm going to keep right on calling her Charlie, just as I've done from the first day I saw her and fed her on the pond. You're a good bird, Charlie!"

At the sound of his voice, the swan looked toward him and nodded her head. If only that other man would go away, she could be more friendly.

April Fool

ON FRIDAY MORNING about nine o'clock, a patrolman walked into the police station and said to the sergeant at the desk, "Hi. The pond is clear of ice. Did you see the swan on it?"

The sergeant looked over to the calendar, where the page exposed showed April 1. "Yeah," he said. "April Fool!"

"No kidding!" the patrolman protested. "Alice is swimming around. There's a dozen people watching from the shore."

"Are you sure there aren't two swans?" the sergeant asked, still unconvinced.

"No, just one. It's the same swan that was there

last year. I guess there was no need to have her indoors any longer."

Alice kept her distance from the road as she explored the millpond, reviving old memories and finding forgotten kinds of greenery growing in the shallows. Her dignity had been hurt again by the same professor. He had pounced upon her in the corner of her sunny house and locked an aluminum band around each ankle. The broad band bore identifying information. He had hauled her to the station wagon, and had held her while a friend drove the car to the pond side. The gray-haired man had come along, but he said very little. He had put into his pocket a couple of the white feathers cut from her wing. He would save them for the naturalist and his wife, who were away on a long trip.

Once in the water, Alice made no attempt to fly. She just swam away as fast and as far as she could. She looked for a quiet bay where no one would come near her. This morning she hadn't even taken time to preen herself. Gradually she realized that she could take a bath. She began sluicing water over her back, rinsing away the dust that had clung to her feathers all winter. She ignored the people who were pointing at her from the road. Most of them had never seen a swan bathe before.

There wasn't much privacy out in the pond. Wherever she went, a photographer kept appearing along the shore, aiming his camera at her. He wanted a picture for the newspaper.

On Saturday and Sunday, hundreds of people came to see Alice. They arrived on bicycles or in cars, to cruise slowly along the road beside the pond. A few walked.

The next day, Alice vanished. Two small boys, who were playing at the edge of the water, knew the secret, however. They told their schoolteacher, "The swan has gone away up the river."

There was no road along the river, only a crooked foot path, explored mostly by small boys. The Swan Master, who felt it was his job to find Alice, had no intention of walking along the crooked path. So he and his son got their skiff out of its storage shed and hauled it over to the water. They attached their little outboard motor and soon had it running. Across the pond they sailed, and up the river, looking for Alice.

Alice heard them coming and swam quickly up a narrow stream that emptied into the river. She hid under some alder shrubs and stayed there most of the day. The Swan Master and his son could not understand where the bird had gone. They knew she couldn't fly. They did see a weasel, still in its winter fur coat of ermine white. It worried them. Would a hungry weasel attack a gentle swan?

The gray-haired man who had fed the swan all winter came trudging along the crooked path in late afternoon, carrying a bag of bread. Every little while he called out, "Here, Charlie, come and get your supper." He had a wonderful New England voice that rang through the woods and echoed down the river. It reassured the swan. After she heard him call a few times, Alice swam out of her hiding place and ate hungrily the slices he tossed to her. She still had a friend, after all.

The Sociable Swan

APRIL PASSED, and May, and June. The air grew warm, the sunshine bright. Leaves appeared, and spread, and cast their shade. Yet the swan rarely ventured from the narrow creek where she could hide.

The townspeople almost forgot that there had been a swan. Scarcely anyone was near to notice when her white feathers began to float from the creek to the river, and into the pond, and across to the dam, and down to the sea. A lot of her small swansdown feathers littered the creek margin, where she often stood to preen herself. No one realized how completely her winter feathers had been replaced by the lighter summer coat. She could stand up high and beat the air with her wings for morning exercises. Now that both wings were fully feathered again, she could even fly if she had to.

Hot weather came in late July. By then the shade of the trees near the end of the little creek kept the swan cool. One warm afternoon, she saw two people walking toward her. It was too early for the man who always called, "Here, Charlie, come and get your supper." So she started up the creek to hide under the alders. She stopped half way, for the sound of their voices was a pleasant friendly sound.

She recognized their silhouettes against the sky beyond the trees at the top of the narrow valley. The naturalist and his wife had returned to her. Excitedly, the wife said, "Look! There she is! Isn't she lovely?"

The swan reacted quickly. She thrust forward her long neck and slanted it upward, calling loudly, "Runh-nh-nh-nh!" Then she raised her great wings and beat the air with them, blowing a regular snowstorm of down feathers from the edge of the creek.

The naturalist turned toward his wife. "She remembers us." Slowly they moved toward the swan, ready to stop and wait if she showed any sign of being frightened.

"Do you really think she recognizes us?" the

wife whispered. "Or has she just become so tame she trusts everyone? I hope she hasn't lost all her natural instincts."

The swan swam over to the side of the creek in the direction of her visitors. She seemed ready to come ashore. But the bank was too steep, although she went one way and then the other looking for a good place to step on land. When her visitors came right to the edge, she lost some of her courage. She swam out a safe distance, but did not go away. She would not reach for a piece of lettuce held toward her. But once it was tossed into the water, she gobbled it hungrily. To her, lettuce was a special treat.

The swan watched her visitors sit on a rock near the creek, and heard them talk to one another in low voices. Every once in a while they tossed a small piece of bread to her. She kept alert to see when the next pieces would come. But soon they had no more to give her.

"I didn't know we'd find the swan, or that she'd come to us after all these months."

The swan was still close to her two friends when the gray-haired man arrived.

"Oh!" he said, in his brusque way, "welcome back. Have you given Charlie all she wants to eat, or do you think she'll take some more? Here, Charlie!" And he flipped a whole slice into the water within easy reach of the swan.

Thereafter the swan had these favorite visitors every day. Rain or shine they came in the morning, and at noon, and once more before the light grew dim. The naturalist and his wife always tempted the swan with lettuce and other interesting things to eat. Soon she grew confident enough to accept food from their outstretched fingers. She enjoyed their company. Whenever they arrived, she swam back and forth close by. Or she nibbled at plants that grew along the shore near them.

On the last day of July, the swan did something she had never done before. After her friends had left, at noon, the naturalist disappeared from view, walking up the hill, going home. But she could still see his wife at the top of the hill.

With a big pair of shears, the naturalist's wife was trimming a thick clump of tall grass that threatened to smother a green carpet of mosses, ferns and lichens. She wanted to see them live in the dappled sunshine that filtered through the leaves of the trees overhead. For almost an hour she clipped and weeded the area, finding all sorts of wildflower seedlings to give special care. She was so intent on what she was doing that she forgot all about the swan. She neither saw nor heard the heavy bird slowly and laboriously climbing the hill.

The swan stood quietly for a while unobserved, just a few yards away. Then she extended her neck, sloping it upward with her beak high, and called, "Runh-nh-nh-nh!" The naturalist's wife jumped and dropped her shears. She turned, saw the swan, and laughed aloud at her own surprise.

"Whatever are you doing up here, Swan-ie, so high on the hill, so far from the water?"

Alice repeated her greeting, and then folded her neck into its customary, graceful S-curve. "Huh-huh!" she said, and nodded just perceptibly.

"I wish I knew what you've come to tell me. I'm sure it's important, or you wouldn't have climbed all the way up here with the message."

Alice bent down and plucked at a piece of moss.

"I don't believe that's what you came for. And if it is, I don't have another bit of food for you." She was trying hard to imagine what the swan wanted. Now her big feathered friend was nibbling at some of the grass she had cut—just playing with it, or showing off how daintily a hard beak could be used.

"I don't know what to do for you, Swan-ie. But I've got to go home. I've finished my weeding. And you're safer back in the water. Maybe later I can find out why you followed me up the hill."

The naturalist's wife walked around the swan and down the hill. Alice waddled after her, and floated out on the creek. "Runh-nh-nh-nh!"

"I'm sorry, Swan-ie, but I'll have to leave you. After lunch I'll come back with more food. Goodbye, until then."

And she disappeared over the hill, picking up her shears as she went.

Home again about ten minutes later, she found her husband looked puzzled.

"I've just had a telephone call from those people who live on the Landing Road, close to the boat dock on the estuary. They say that a swan is swimming down there among the sailboats. Do you think it's another swan, or is Alice leaving us?"

The naturalist's wife gasped, then told her husband exactly how the bird had climbed the hill to her. "Was that what she was trying to tell me? Was she saying goodbye?"

Hamilton

Alice could fly again. Yet for several hours she swam about in the estuary among the sailboats at their moorings. She tasted the salt of the water, and found it stimulating.

The busy Swan Master heard that Alice was headed out to sea. He hurried down to the estuary, arriving just in time to see the swan spread her great white wings and take off. She flew low over the water, eastward around the bend and out of sight.

Now what should be done? The professor had come back to town the night before, bringing a mate. Already they had named the new bird Hamilton. But what good was Hamilton without Alice?

He waited a while, hoping she would return. He talked with some other people who had watched Alice go. Then he went to call the professor who was looking after the new swan.

"Isn't that fate, for you!" exclaimed the professor. "Every time we get a mate for Alice, something happens!"

Hamilton was a wild swan, who came legally from a nearby New England state. He still had not reached full size, although his first coat of white feathers would soon be replaced by a new set. He would need the warmer ones to protect him from the winter cold. He was such a skinny bird that the professor worried that he might have a swan disease, or perhaps a tapeworm or two. He felt it would not be wise to release Hamilton on the pond until the expert from the animal science department made a thorough check. And the expert was away on a fishing trip. Hamilton would have to remain in a pen until the expert returned and made the necessary tests. It might be several weeks. There was no way to hurry, and no use in taking chances on giving Alice a disease—if she came back.

Late in the afternoon, Alice flew westward again.

She came winging high in the air over all the sailboats. She passed above the highway and the dam. Exhilerated by her own flight, she continued past the millpond and up the long reach of the river to its very end. There she almost crashed into a tree, making a U-turn. But she managed and swept back along her course. Just as she reached the end of the little creek where she had hidden out for so much of the summer, she splashed down. She wagged her tail, adjusted her wings, and swam over toward the shore. In a few minutes her gray-haired friend would arrive, calling, "Here, Charlie, come and get your supper!"

The next morning, Alice flew up the river again, made her U-turn, and sped all the way to the estuary. Her wing muscles needed exercise. Soon she could go wherever she wanted to without tiring.

She settled among the sailboats and found some tasty seaweeds close to shore. She enjoyed them without knowing that she needed them to keep her healthy. They contained important amounts of iodine, an essential mineral that was rare in the pond plants. She was still choosing the best pieces to swallow when a fisherman came up the estuary in his motor boat. The sound of his motor disturbed Alice, and she swam toward the dam as fast as she could go. In her hurry she forgot that she could fly.

The tide was low, and the water in the estuary did not cover the stones and boulders in the broad basin below the dam. Hobbling as best she could, Alice got herself over and between the rounded rocks. Finally she floated on a narrow band of deeper water right below the lip of the dam. When the tide returned, she could reach the estuary easily and go wherever she wished.

The Swan Master came home for lunch and saw the swan below the dam. Quickly he telephoned to the professor. "Alice is in a place where we could catch her easily," he said. "I think we should clip one wing and put her on the pond. Then we can be sure she'll be there when Hamilton gets his health certificate and we can release him.

"I'm not very anxious to handle Alice again," said the professor. "And a lot of people in town won't like it if we keep her from flying. She came of her own will, and she should be able to leave if she wants to. Her freedom to choose is part of what gives her appeal. But you're the Swan Master. I'll come with my hip boots if you insist."

"We've got to have Alice here to welcome Hamilton. The town is counting on having a pair of swans on the pond. I believe we have no choice. We'll have to stop Alice from flying away."

A New Friend in a Secluded Backwater

ALICE FELT anything but free when the men released her on the millpond. She swam as fast as she could go, away from the people, away from the dam, away from the pond, away . . . away! That same professor had taken advantage of her again.

All afternoon she swam. As night came, she was still going furiously, far up the long reach of the river. She reached a broader place where the water was choked with vegetation, and couldn't see her way in the dark. So she waited, unable to go on, unable to sleep in the unfamiliar spot, until the dawn. Then she started on again, pushing her body between and over the soft green plants to open water beyond them. Onward she went, to the very end of the channel. Dodging the low branches, she got herself far into a dense thicket of alders growing out of the water. There she slept, exhausted from her travels.

Late in the afternoon, she head the voice of the gray-haired man who called her "Charlie." He kept calling, calling, calling. But she didn't budge. She was tired. He was tired too, for he had trudged the whole length of the crooked path four times, looking for the swan. He felt sure she was hiding somewhere among the alders that overhung the river, but he did not find her.

Other sounds kept the swan nervous. A family of muskrats rustled as they went back and forth through the cattails where their house was hidden. The swan jumped a little whenever a muskrat sneezed. She drew back too when she heard the alder branches creak together as two deer came down to the water to drink. In the dim light of evening, she could hardly see them although they stood so close.

A gray fox came out of his den under a fallen tree and barked several times, almost like a dog. Then he walked silently along a trail he knew and hunted mice. He was an animal of the night, and for him this was breakfast time. His sensitive nose told him that the swan was hiding among the alders. But he did not recognize the odor as something good to eat, and he passed her by with only a sniff in her direction. She heard him sniff and tensed her muscles until he went away.

Alice dozed in the darkness. Suddenly she awakened, aware of a vibration in the water, shaking her feet ever so slightly. A minute later, she heard a strange flubbing sound, and gutteral grunting noises, and then the click of teeth. An otter had caught a big perch under water, dragged it struggling to the shore, and now was sharing it with his mate. Nothing suited them so well as a midnight snack of fresh fish. When they had cleaned up every bone, the two otters climbed the bank and scampered one after the other to a muddy slope they had enjoyed many times before. The otter who had caught the fish was still dripping wet. He ran to the top of the slope and slid down it, making it more slippery than ever. Splash! He went into the water. Splash came his mate right behind them! Quickly the two of them paddled to shore, stepped out and shook themselves, and ran to the top of the slide again.

The swan quivered every time she heard that splash. For almost half an hour the scary sounds kept coming, until the otters grew tired of their game and went home to sleep.

Soon after daylight began to spread across the sky, Alice was startled by a new sound. With a loud whuffing, like somebody shaking a wet bedsheet, a great blue heron braked his forward flight. He let himself down into the marshy part of the river close to the alders where Alice was hiding. His narrower wings were almost as long as hers. His legs were longer, and held him high above the water. He folded his slender neck into a flattened S-curve, which brought his long sharp beak down against the gray-blue feathers on his breast. Already his sharp eyes were searching for a fish or a frog he could spear for breakfast.

From close to shore where the alders had their roots, about three dozen beetles swarmed out over the water. Each kept its back dry while it drove its oval, flattened body forward with rapid paddling movements of its six short legs. Every one of these whirligig beetles had the same jet-black armor shining toward the sky. It held its feelers ahead of it and sensed every vibration in the water film. With information from those feelers, its tiny brain acted like a computer, telling the whirligig when to turn and which direction to move in to avoid colliding with other things. Although these insects swam at high speed, they never bumped their neighbors or hit a floating twig. They darted ahead, though, to inspect anything that fell onto the surface film of the water or floated up from below. In this way they found their food.

Like a flotilla of miniature boats, the whirligigs darted past the swan. They scattered in panic when the heron struck downward to seize a little fish. But they regrouped at once and were on their way. After a few minutes, the swan followed them into the open and began probing for food in the shallow water.

It took Alice several days to get used to her wild neighbors in the secluded bay at the very end of the river. Gradually she learned that she had nothing to fear from any of them. The great blue heron complained at first when she swam past him. He half raised his wings and spread them to make himself look bigger. He drew back his head and uttered a hoarse croaking sound as a warning that she was to come no closer. No fisherman likes to have the fish disturbed by something swimming over them. Fish are hard enough to catch under the best of circumstances.

Alice did not realize that anyone owned the secluded bay, or that she had been seen hiding there.

But the naturalist knew where she was, and so did the professor who had clipped her wing. One morning she saw the two of them coming. She rushed into the alder thicket where she hid when anything disturbed her.

She peered at the men through the alder branches as they came to the edge of the water at a point about a dozen yards away. There they gently set down a white bundle, which had a long projection that looked amazingly like the neck and head of a swan. They rolled the bundle over and untied a cloth. Two black feet, very much like her own, showed as soon as they took the cloth away. They rolled the bundle back until the feet were under it, and untied another cloth. Suddenly wings appeared. The men eased the bundle into the water and gave it a little push. It moved away from land, and kept on going. It was Hamilton!

He was skinnier than ever. He had his health certificate, but he was too weak to fly. He just swam slowly. Then he dipped his beak into the water and sipped some. It had a different flavor, much better than the clean treated water he had had in the pen. He bumped into some green plants and nibbled at them. He had almost forgotten how water plants tasted. They were exactly what he needed. He ate and ate.

After a while, Hamilton was full of water plants. Then he took time to gaze around at the strange pond where all the good food was. He saw the professor and the naturalist sitting perfectly still on the shore. And then he noticed something white moving under the alder branches. He turned around and peered more carefully. He had discovered Alice. Except that she was so much more beautiful, she looked just like himself. He had company—a bird of his own kind. He swam a little closer.

Alice had not seen another swan for almost a year. She shoved through the alders and swam out into the open to look at the scrawny bird. His feathers were dirty. His neck was thin. The knob on his beak scarcely showed. He seemed to have no energy. He wasn't much better than no swan at all! Alice turned away and paddled across the bay to find something to eat. Hamilton followed her at a respectful distance. He stood guard while she had her lunch.

Because of a Swan

WITH TWO SWANS feasting in the secluded part of the river, many of their favorite plants became scarce. No longer did masses of green vegetation clog the backwaters. The birds could swim easily almost anywhere.

Several times, Hamilton started down the long reach toward the dam. But Alice would not follow him in that direction, and he always turned back to stay with her. A month of wild food in abundance had been like a tonic. He was heavier. His feathers shone, clean and white. New ones were growing in—his winter coat. Soon he would be able to fly and lead Alice to some salt marsh along the coast where they could spend the winter—if Alice would ever move.

One day Hamilton grew impatient. He set out by himself and swam all the way to the millpond. There he met a pair of coots. They were silly birds, with fat bodies, gray feathers, and ivory-colored beaks. The coots kept dodging after one another among the cattails, cackling to one another like chickens. They had stopped at the millpond on their way to the Everglades in Florida, and stayed a while because there was so much food. Hamilton filled his own stomach. Then he swam back to get Alice.

Maybe she missed him. She stayed closer than usual all night. And when once more he headed toward the millpond, she went too. Toward dusk the two swans entered the pond. They spent the dark hours with their heads tucked into the feathers on their backs, while they stood on the little low island out in the middle.

At last the millpond had a pair of swans. As they swam about, it was easy to know them apart. Alice held her neck in a graceful S-curve, and she showed her greater confidence by elevating her wings slightly above her back. She was "busking" her wings, as Scottish people say, and she busked them a little extra whenever she wanted to show off. Hamilton kept the lower half of his neck almost straight, upright or leaning forward or back. He curved the upper half, and appeared slightly more dignified and formal. He scarcely ever busked his wings. That gesture was for early spring, when the season for courtship came or he had territory to defend.

While Hamilton was close, the gulls never bothered Alice the way they once had. They kept to a greater distance and used their energies in quarreling among themselves. In his company she felt secure in going anywhere around the pond. He acted like a guardian, without interfering with whatever she wanted to do.

Hamilton distracted Alice from her human friends. His presence, in fact, forced her to decide that she was a bird, not a person. She had begun to mistake her own identity because so many people had been feeding her.

At first Hamilton was merely company—a bird of her own kind who had dropped in for a visit on her private end of the river. But as he had gained strength, his importance had also grown. He was always there beside her, respectful and dignified. The black knob above his beak was getting bigger. At times he looked splendid. He could mature to be magnificent. Alice busked her wings in a proud arch.

Imperceptibly a friendship ripened. A bond grew between the two. They were not yet mates; that would have to wait. With springtime could come the courtship, the ceremony joining the two in partnership for life. The future of the swans had just begun. A century of companionship lay ahead of them. There was no hurry.

Birds flying high above the pond noticed the white swans and came spiralling down to rest and feed. Four baldpates paused for several hours and then took off to continue their southward migration. A flock of scaup ducks, which had spent the summer diving for mussels off the coast, dropped down in wedge formation. They ate a few green plants, then pattered off across the water until they gained speed, showing their broad white wing-stripes. A few grebes stayed for more than a week.

For all of these birds, the swans advertised a safe sanctuary. And the more ducks that came, the more others joined them. Like magic, the swan's presence touched upon the lives of countless other water creatures. They shared their security with everything wild that could use the pond.

Because of one swan, there were now two. And because of them, there were many other birds. The sociable pair and their reflections in the water added a new quality to a scene so familiar that its beauty had been largely forgotten. Because of two swans, the millpond and the river above it gained more importance than they had had at any time since the old dam was built.

Books for Additional Reading & Reference

AUSTIN, O. L. Jr. *Birds of the World* (New York: Golden Press; 1961) 316 pp.

DELACOUR, JEAN, AND PETER SCOTT *The Waterfowl of the World* (London: Country Life; 1954) vol. 1 includes swans.

GILLIARD, E. T. *Living Birds of the World* (Garden City, N.Y.: Doubleday & Co.; 1958) 400 pp.

PACA, L. G. *The Royal Birds* (New York: St. Martin's Press 1963) xi + 164 pp.

TICEHURST, N. F. *The Mute Swan in England: Its History, and the Ancient Custom of Swan Keeping* (London: Cleaver-Hume Press, Ltd.; 1957) xiii + 131 pp.

LORUS J. MILNE and MARGERY MILNE, professors and exploring scientists, have worked together since their college days. They have lectured at a number of universities together, and their travels have taken them over 300,000 miles through four continents. Photographs they have made have appeared in many newspapers and magazines here and abroad.

They are the authors of many books, most of them for adults, also of two other books for children, *Because of a Tree* and *The Crab that Crawled Out of the Past.* Articles they have written have appeared in many national magazines.

LORUS J. MILNE was born in Toronto, Canada, and is now a citizen of the United States. He attended the University of Toronto, where he was named a Gold Medalist for his work in biology, and received a Ph.D. from Harvard University.

MARGERY MILNE, a native of New York City, attended Columbia University and received her M.A. and Ph.D. from Radcliffe College.

The Milnes live in Durham, New Hampshire.